RELATIVE INTIMACIES

AF623096

INTERSUBJECTIVITY VOL. III

RELATIVE INTIMACIES

Sternberg Press

Edited by Lou Cantor and Emily Watlington

This book was made possible through the kindness and support of numerous collaborators. Deepest thanks goes to Thomas Hug and the Foundation pour les Arts Visuels, who have supported the vision that drives this book series from conception to completion. Katherine Rochester's insight was essential for our conceptualization. To our deft, astute, and generous contributors: thank you for entrusting us with your work and for your diligent care with your texts, especially during this difficult year. We are also grateful to Aura Rosenberg and Josie Thaddeus-Johns for support and inspiration.

artgenève
artmonte-carlo

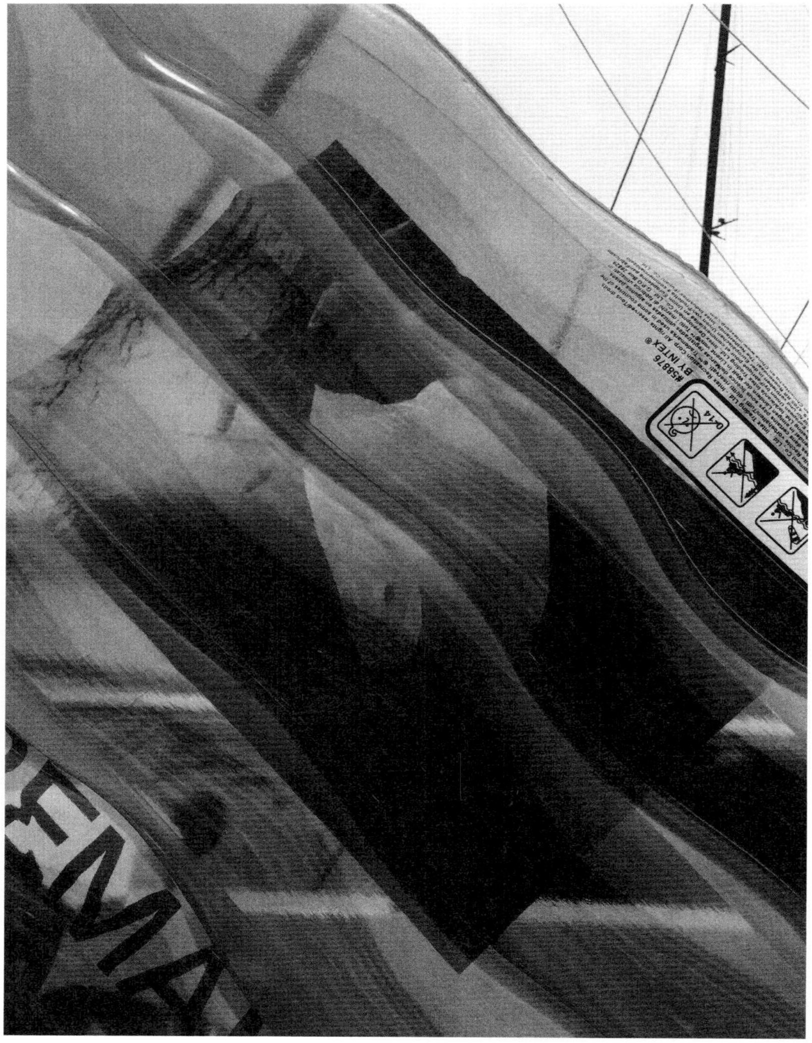

The Detox Paradox

December 2018. Jack Dorsey, CEO of Twitter, tweets announcing his withdrawal from the world.[1] He has journeyed to Pyin Oo Lwin, Myanmar (Burma), in a classic Silicon Valley pilgrimage to the exotic heart of new age spirituality. Dorsey has long positioned himself as a spiritual influencer for contemporary technoculture, proselytizing about the need for humanism, intimacy, and nature to balance the networks of outrage that he himself has helped create.

And so, Jack Dorsey drinks salt juice each morning. He fasts. He takes ice baths. While in Burma, he undergoes a week of silent meditation, detoxing himself from the excesses of technology. But the detox itself is rationalized in terms of technological optimization. Dorsey cites vipassanā, an ancient Buddhist practice that was itself rearticulated as a popular modern movement partly through American practitioners. For Dorsey, vipassanā is about "hack[ing] the deepest layer of the mind and reprogram[ming] it."[2] He sees Buddha as a proto-biohacker, characterized by his "rigorous scientific self-experimentation."[3] Dorsey positions himself as an extension of Buddha's noble science, enabled by the latest smart machines. Seated in a Mandalay cave (but one laid out with a tiled floor), he tracks himself with wearable devices, monitoring his heart rate, his calories, his REM sleep.

Outside the cave, Burma remains less than tranquil. In the months preceding Dorsey's trip, hundreds of thousands of Rohingya Muslims had fled the country, seeking relief from systematic persecution by the military regime. But should Dorsey have been more mindful of the wider ethical context? One commenter thought not: "[Dorsey] came here just to do something during his birthday." He was just trying to cleanse himself from the swamp of data. Allegations of ethnic cleansing were a problem for someone else, for some other day.

The detox is now an established stage in the ouroboros of technology consumption. First, we subsume ourselves in a data surround, in which the sheer overproduction of data becomes normalized as the rhythmic backdrop to everyday life. Technoculture then offers a new range of data analytics that we might further consume to make our (always temporary) escape. Some tools merely visualize and collate the extent of our technological malaise; others turn to techniques like Pavlovian conditioning to literally shock us out of it. Two MIT students describe the almost nonconscious habit through which they end up wasting time on Facebook—it is as if, they say, they are "dragged there by some mysterious Ouija-esque compulsion."[4] Their solution is the Pavlov Poke: a browser activity–monitoring system tied to a circuit board, that sends electric shocks to the user upon detecting nonproductive activity. Detoxing restores control—but only if you agree to suspend your ordinary boundaries and accept heretofore exceptional forms of technological intervention.

1 Jack Dorsey (@jack), "For my birthday this year, I did a 10-day silent vipassana meditation, this time in Pyin Oo Lwin, Myanmar. We went into silence on the night of my birthday, the 19th. Here's what I know," Twitter, December 8, 2018, 7:19 p.m., https://twitter.com/jack/status/1071575088695140353.

2 Jack Dorsey (@jack), "Vipassana's singular objective is to hack the deepest layer of the mind and reprogram it: instead of unconsciously reacting to feelings of pain or pleasure," Twitter, December 8, 2018, 7:19 p.m., https://twitter.com/jack/status/1071575097197002752.

3 Jack Dorsey (@jack), "Vipassana is a technique and practice to 'know thyself.' Understanding the inner nature as a way to understand...everything," Twitter, December 8, 2018, 7:19 p.m., https://twitter.com/jack/status/1071575094013521923.

4 Adi Robertson, "An Electroshock Facebook Deterrent Converts Mindless Browsing into Pain," *The Verge*, August 23, 2013, https://www.theverge.com/2013/8/23/4651366/pavlov-poke-treats-facebook-addiction-with-shock-aversion-therapy.

5 Susan Maushart, *The Winter of Our Disconnect: How Three Totally Wired Teenagers (and a Mother Who Slept with Her IPhone) Pulled the Plug on Their Technology and Lived to Tell the Tale* (New York: Penguin, 2010), vi–vii, quoted in Trine Syvertsen and Gunn Enli, "Digital Detox: Media Resistance and the Promise of Authenticity," *Convergence: The International Journal of Research into New Media Technologies* 26, no. 5–6 (2020): 1276.

6 Sandro Formica, *Digital Detox: 7 Steps to Find Your Inner Balance* (self-pub., 2015), quoted in Trine Syvertsen and Gunn Enli, "Digital Detox: Media Resistance and the Promise of Authenticity," *Convergence: The International Journal of Research into New Media Technologies* 26, no. 5–6 (2020): 1276.

But what are we trying to fix about ourselves when we detox? The fast-growing detox business deftly leverages the language of the self-help and wellness industries, commodifying authenticity as an object of performative nostalgia. Popular detox manuals describe smart machines as a "force field" separating humans from each other and from themselves;[5] the detox, in turn, will reconnect us with our bodies, as well as with nature and all its "healing power."[6] Detox apps also draw freely from the iconography of addiction and mental health; one popular app (AppDetox) invites users to "Get your dose of AppDetox now," complete with a surgical needle for the app icon.[7]

All this, of course, means more labor and responsibility for the hyperactive subject, parallel to the ways in which productivity technologies in the workplace create new demands on the worker to manage themselves.[8] Detox apps like Moment typically provide dashboards for smartphone use data, such as screen time versus "quality offline time."[9] As such data become integrated into the base operating system (such as Android's Digital Wellbeing, or iOS's Screen Time), they ironically feed into more notifications and alerts *about* the detox as an object of quantified optimization. Meanwhile, social media executives, nestled in their boundless riches, confess they try to stay off social media and limit their own children's screen time. The reader is thus asked to reflect on their own moral failure to individually resist an industrialized compulsion. Like the physical detox in the realms of food and health, in which I *owe it to myself* to regularly cleanse my body of harmful substances, the digital detox makes a moral responsibility of the need to purify, sanitize, empty out our bodies and minds.[10] Here we might also think of enduring ideas around digital hygiene:[11] the growing work of daily maintenance the user must perform to stay clean, stay safe online, whether from viral infections or anti-vaxxer disinformation.[12]

As with Dorsey's meditations, the digital detox industry's gestures towards the holistic and the intangible are articulated through a highly calculative, optimising mode of seeing. Detox characterizes its enemy as time wasted, life frittered away on the screen, which can and must be corrected through an almost surgical separation of data's harms from its benefits. We hope to "reconnect" with ourselves and build a happy life out of objective calculations, without the consequences of side effects like distraction and addiction—which are construed as not so much failure on the part of the technology, but "diseases of the will."[13] Yet the process of managing the growing array of smart machines and the data they track entails returning to the treadmill of data-driven optimization. Philosopher Byung-Chul Han calls this condition *Müdigkeitsgesellschaft*, or "fatigue society": that stagnating malaise in which the to-do list is always growing, the inbox is ever dinging, the Twitter feed endlessly repopulating, all in ostensible aid of a

7 Hoornaz Keshavarzian, "Digital Detox Hegemony: The Misconception of Offloading Technology Fatigue" (unpublished, Simon Fraser University, 2020).

8 Melissa Gregg, *Counterproductive: Time Management in the Knowledge Economy* (Durham, NC: Duke University Press, 2018).

9 "Moment: Cut Screen Time" (Moment Health, Inc., 2014), iOS, last modified March 2, 2021, https://apps.apple.com/us/app/moment-cut-screen-time/id771541926.

10 For the connection between food metaphors and digital detoxes, see Theodora Sutton, "Disconnect to Reconnect: The Food/Technology Metaphor in Digital Detoxing," *First Monday* 22, no. 6 (June 2017), https://journals.uic.edu/ojs/index.php/fm/article/view/7561.

11 Tung-Hui Hu, *A Prehistory of the Cloud* (Cambridge, MA: MIT Press, 2015), 57.

12 For one such analysis, see Cait McKinney and Dylan Mulvin, "Bugs: Rethinking the History of Computing," *Communication, Culture and Critique* 12, no. 4 (2019): 476–98.

13 See Mariana Valverde, *Diseases of the Will: Alcohol and the Dilemmas of Freedom* (Cambridge: Cambridge University Press, 1998).

subject for whom the overabundance of activity is valorized as essential virtue.[14] Dorsey's meticulously planned detox, gilded with a touch of disaster tourism, seeks a reset button on the human machine. The detox purges the toxic build-up of the excess information and stimulation that characterizes the smart society—such that Dorsey may return to the fray, all the more energized, for business as usual.

We detox from our machines only in order to return to them as more efficient and productive subjects. We detox according to the very logic of what we detox from: that of optimization. These contradictions reflect a broader tension in what I have called elsewhere "data's intimacy," or, the promise that emerging "smart" machines will help you know yourself better than ever, that autonomous surveillance and predictive analytics will provide objective knowledge of ourselves.[15] Across wearable devices, smart home assistants, and other fixtures of the Internet of Things, we find a new ecosystem of machines that cling to your body, listen in from inside the home, and are sometimes literally under your skin. They observe, record, and, in doing so, articulate your existence in terms of calculable proxies.

But intimacy is never a one-way street. The word often evokes a sense of human togetherness and transmission. Yet its roots in *intimus,* meaning the inmost parts of the self, tells us that this proximity can apply elsewhere and in different ways. Our understanding of intimacy as love, friendship, hanging out, liking, following, laughing reminds us that even as we hope that intimacy makes us whole, it can also be a disruptive and fraught process that opens us to vulnerability. Intimacy is not just about getting "in touch" with our inner selves; it also turns us inside out, exposes us, leaves us vulnerable. Forging new connections often entails creating new boundaries as well, demarcating good kinds of proximity and contagion from bad ones. How "close" should we keep our smart machine that wants to live in our beds and extract data from our skin? What other kinds of connectivity, exposure, visibility come hand in hand with the kind of intimacy proposed by the smart machine?

Involuntary Vulnerabilities

In the early days of the popular web, many wondered if the reduction of certain sensory channels—from real faces to emoticons, real-time to asynchronous conversation—would impoverish human intimacy. Just as consequential, it turns out, are the ways in which algorithmic systems regulate the kinds of people, databases, opportunities, and communities with which we gain meaningful contact. In *Purity and Danger,* the anthropologist Mary Douglas shows us how cultural conceptions of cleanliness and (im)purity regulate acceptable forms of proximity.[16]

14 Byung-Chul Han, *Müdigkeitsgesellschaft* (Berlin: Matthes & Seitz, 2010).

15 Sun-ha Hong, *Technologies of Speculation: The Limits of Knowledge in a Data-Driven Society* (New York: New York University Press, 2020).

16 Mary Douglas, *Purity and Danger: An Analysis of Concepts of Pollution and Taboo* (London: Routledge, 1984).

17 For more on this subject, see John Cheney-Lippold, *We Are Data: Algorithms and the Making of Our Digital Selves* (New York: New York University Press, 2017); and Luke Stark, "Algorithmic Psychometrics and the Scalable Subject," *Social Studies of Science* 48, no. 2 (2018): 204–31.

18 N. Katherine Hayles, "Cognitive Assemblages: Technical Agency and Human Interactions," *Critical Inquiry* 43, no. 1 (Autumn 2016): 32–55.

19 Kenza Bryan, Sian Griffiths, and Jon Ungoed-Thomas, "Revealed: Betting Firms Use Schools Data on 28m Children," *The Times (UK)*, January 19, 2020, https://www.thetimes.co.uk/edition/news/revealed-betting-firms-use-schools-data-on-28m-children-dn37nwgd5.

One striking aspect of the so-called smart society is the degree to which those demarcations are enacted, black-boxed, and concealed through technologies of intimacy. Here, the human proximity and connection afforded through the smart device or the social media platform is the *side effect*; what is primary is the industry's ability to collect user data from bedrooms, bathrooms, and every crevice of what was once considered private or domestic. It is a question of how our moods or sleep patterns, exercise routines or sex lives can be made most compatible to the array of available sensors and analytical processes such that they can be rationalized into "insights" and predictions. It is intimacy defined not by a robust theory of the personal, private, or human, but by the degree to which scalable and automated systems of quantified prediction can *get close to* the most personal and mundane aspects of individual subjects.[17]

In short, the advent of smart technologies—made in the name of human intimacy, a means to connect with "our true selves"—is also a project for our involuntary vulnerability to machinic measurement and to the quantifiable standards of health and wellness, productivity and efficiency that they entail. N. Katherine Hayles calls this "somatic surveillance," or the exteriorization of what previously remained relatively nonrational and nonconscious, in order to avail it for calculative rationalization.[18] Here, the kind of proximity that matters for profit and predictivity pays little heed to the kind of intimacy appreciable

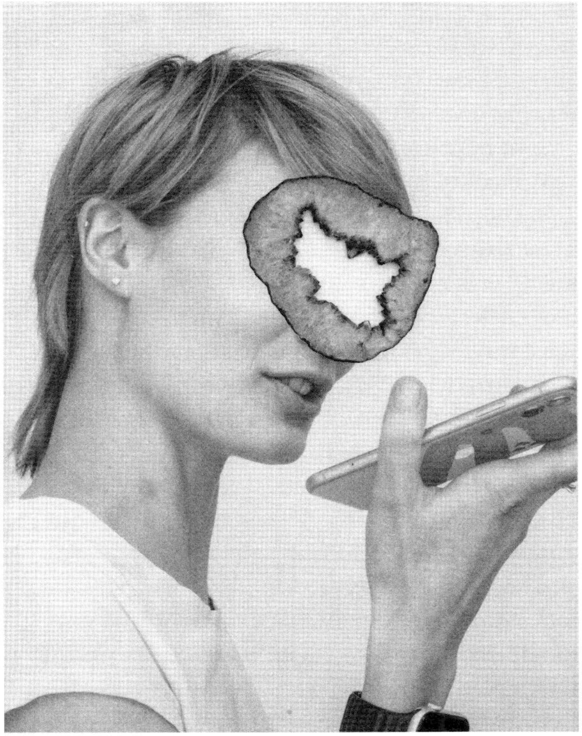

Lou Cantor, *Apocalypse not by Abolition but by a Transfiguration*, 2021. UV print on aluminium honeycomb plate, amethyst

by human sensory experience. Behavioral data is collected from schoolchildren originally in the name of community safety but is then quickly passed on to companies—in one case, betting firms—that can use the data to identify potential customers.[19] Our faces, that long privileged point of human intimacy, are first voluntarily uploaded for social connections but subsequently recombined to train facial recognition datasets. Eventually, your images become more promiscuous than you could ever be, becoming caught up in datasets and institutions well beyond any practical ability of users to give meaningful consent. Yahoo's YFCC100M dataset, released in 2014, included faces scraped en masse from Flickr archives as old as 2004.[20] Faces photographed and shared in one's adolescence might still circulate a decade on—totally unbeknownst to their owners. None of us know now where we will next encounter our own faces or where our faces are being read and measured. All over the world, faces of ordinary individuals are already being used to bill customers, as in India's Chaayos coffee chain; to admit (and bar) visitors, as in China's Hanzghou Safari Park; to scan passengers on airplanes, as with America's Delta Airlines. Trevor Paglen notes that the vast majority of images today are produced by and for machine communication.[21] The same is true for personal data. What we actively communicate to a platform like Facebook through our words and likes is dwarfed by more passive, invisible forms of communication, such as the company's ability to invisibly track online activity outside the platform through pixel tracking. Such data is pumped and circulated to fatten datasets that are often black-boxed out of human scrutiny and understanding.

Technologies of datafication are designed to share and leak data; such promiscuity is crucial to the economic interests of the tech industry as we know it.[22] For it is precisely in the promise of collecting data for one reason and using it for another, in recombining data for unexpected new purposes, that the technology is argued to provide extra value.[23] Consider Clearview AI, the now notorious surveillance service that seeks to provide the largest possible database of faces scraped from the internet. Promotional emails from Clearview encouraged its clients to be as voracious as possible, never mind the privacy risks:

> Have you tried taking a selfie with Clearview yet? See what comes up! [...] Try your friends or family. Or a celebrity like Joe Montana or George Clooney.
>
> Your Clearview account has unlimited searches. So feel free to run wild with your searches.[24]

The list of Clearview clients is confidential, though leaks indicate that the product, publicly described as strictly for law enforcement purposes, is also being used by the likes of Walmart and a Massachusetts high school.[25] The human intimacy promised by technologies of datafication, tinged with romantic promises of individual agency and spiritual harmony, functions as a foil for an altogether different logic of proximity between corporations and my data, law enforcement and my body.

20 Bart Thomee et al., "YFCC100M: The New Data in Multimedia Research," *Communications of the ACM* 59, no. 2, February 2016, 64–73, https://cacm.acm.org/magazines/2016/2/197425-yfcc100m/fulltext.

21 Trevor Paglen, "Invisible Images (Your Pictures Are Looking at You)," *The New Inquiry*, December 8, 2016, https://thenewinquiry.com/invisible-images-your-pictures-are-looking-at-you.

22 For more on this subject, see Wendy Hui Kyong Chun, *Updating to Remain the Same: Habitual New Media* (Cambridge, MA: MIT Press, 2015).

23 Michael Betancourt, "Immaterial Value and Scarcity in Digital Capitalism," *CTheory* (June 10, 2010), http://www.ctheory.net/articles.aspx?id=652.

24 Ryan Mac, Caroline Haskins, and Logan McDonald, "Clearview AI Once Told Cops to 'Run Wild' with Its Facial Recognition Tool. It's Now Facing Legal Challenges," *BuzzFeed News*, January 28, 2020, https://www.buzzfeednews.com/article/ryanmac/clearview-ai-cops-run-wild-facial-recognition-lawsuits.

25 Ryan Mac, Caroline Haskins, and Logan McDonald, "Clearview's Facial Recognition App Has Been Used by the Justice Department, ICE, Macy's, Walmart, and the NBA," *BuzzFeed News*, February 27, 2020, https://www.buzzfeednews.com/article/ryanmac/clearview-ai-fbi-ice-global-law-enforcement.

26 José van Dijck, *The Culture of Connectivity: A Critical History of Social Media* (Oxford: Oxford University Press, 2013).

27 Baveja Sarabjit Singh, Dalvi Nilesh, and Sarma Anish Das, Determining trustworthiness and compatibility of a person, US Patent 9,070,088 B1, filed September 16, 2014, and issued June 30, 2015.

28 Singh, Nilesh, and Das, Determining trustworthiness and compatibility of a person, col. 9 and 2.

Speaking of social media platforms, José van Dijck noted that the promise of human connectedness is used to advance greater connectivity at the level of data-driven surveillance.[26] It is not merely that machinic proximity sneaks in, like a thief in the night, under cover of human intimacy. Rather, the very terms by which we seek human intimacy is increasingly *rationalized* in terms of predictive surveillance. Smart technologies achieved widespread buy-in through deeply discounted offers of wonderfully novel connectivity: ride my car, sleep in my home, make friends along the way, escape the soullessness of the mass-produced travel experience. Once we're strapped in for the ride, though, there is a gradual creep in terms of admission. It's not just that Uber rides grow more expensive, shedding the billions of investor funding that had artificially depressed prices in its early years; it is also the new forms of screening and judgment required to play this new game of proximity. Airbnb is now exploring algorithmic screening systems for tenants, where data on individuals might be scraped and collated in an attempt to derive predictions of their "trustworthiness."[27] When machines trawl the web for data like "swear words or phrases that express negative emotion" to derive personality traits like "neuroticism, narcissism, Machiavellianism, and psychopathy"[28]—and in turn use that data to provide or withhold accommodation services—we start to see a new set of gatekeeping functions, new rules of proximity and separation, derived from what was allegedly a technology of intimacy. Giggle, a new girls-only social network, wants to offer the safety of curated intimacy—as long as you undergo scanning through a "bio-metric gender verification software" to show you are really a woman. "Unfortunately, it doesn't verify trans-girls," we are told.[29]

These emerging forms of technological governance extend out of the screen and onto the material conditions of those technologies as well. Smartphone users download PooKeeper to track their own shit, mindful of the rising responsibility to measure and optimize every facet of their personal lives. Meanwhile, Facebook's overworked moderators—tasked with the unenviable work of cleaning metaphorical shit off the platform at high speed—are instructed to log their own toilet breaks, creating ever more fine-grained measures of workplace "productivity."[30] Amazon has long been accused of doing the same with its warehouse workers. Meanwhile, at least one Uber office has been seen segregating its drivers and full employees into separate bathrooms.[31] Who gets to come in contact with whom? Who gets to measure themselves—and who cannot afford to avoid being measured by others? As one early pioneer of self-tracking technologies would later muse, "It costs a lot not to have internet access. I wonder in the future whether people will be able to afford not living with the internet. [...] Not being connected will constitute a new disability."[32] Initially, the

29 "Why is bio-metric gender verification software used to access giggle?," Frequently Asked Questions, giggle, archived February 6, 2020, https://web.archive.org/web/20200206174236/https://joinagiggle.com/faq. The wording of the FAQ was changed on February 8, 2020, following public outcry; see for example Zoe Schiffer, "This Girls-Only App Uses AI to Screen a User's Gender—What Could Go Wrong?," *Verge*, February 7, 2020, https://www.theverge.com/2020/2/7/21128236/gender-app-giggle-women-ai-screen-trans-social.

30 Dave Gilbert, "Facebook Is Forcing Its Moderators to Log Every Second of Their Days—Even in the Bathroom," *Vice*, January 9, 2020, https://www.vice.com/en_us/article/z3beea/facebook-moderators-lawsuit-ptsd-trauma-tracking-bathroom-breaks.

31 Alex Horton, "An Uber Office Segregated Bathrooms for Drivers and 'Employees,'" *Washington Post*, December 5, 2019, https://www.washingtonpost.com/technology/2019/12/05/uber-office-bathrooms-segregated.

32 Chris Dancy, interview by Phoebe V. Moore, in Moore, *The Quantified Self in Precarity: Work, Technology and What Counts* (Abingdon: Routledge, 2018), 191.

question is simple: Do you want a smart assistant that listens in on you? Over time, the equation shifts: Can you afford to refuse the smart assistant if your job, your insurance, your social network effectively require it?

The expansion of involuntary vulnerabilities continues at a rapid pace. Google is actively pursuing partnerships with hospitals and health data companies in order to gather more comprehensive data about your health, your medication, your symptoms—and so is Facebook.[33] Amazon is steadily building its ecosystem of smart home devices for data collection. Its Alexa smart assistant can now ask questions and initiate conversation, and has passed Health Insurance Portability and Accountability Act (HIPAA) certification to handle patient data. Meanwhile, Amazon's Halo fitness band seeks to "create 3D scans for body fat and [listen] for the emotion in your voice."[34] Each possible site of data's intimacy becomes a space for strategies of "accumulation by dispossession,"[35] in which something about us is systematically devalued *such that* it provides surplus value for its canny extractors. Smart technology, as one Silicon Valley marketing director confides, is "all push, not pull. [...] The Valley has decided that this has to be the next big thing so that firms here can grow."[36]

But the push for profit and growth also pushes along—and pushes away—many other kinds of social change in its uncaring stride. The pursuit of intimacy with myself, with others, with the world is conducted through the belief in technology as a shortcut, information as a neutral equalizer, data as purifying element for the messiness of intimacy. At the same time, such intimacy is opening up new paths for the extraction and exploitation of the personal, the private, the domestic, creating new and involuntary points of vulnerability. And whenever you need a break from all that? The detox is here for you, too—if you can afford it.

33 Rob Copeland, Dana Mattioli, and Melanie Evans, "Inside Google's Quest for Millions of Medical Records," *Wall Street Journal*, January 11, 2020, https://www.wsj.com/articles/paging-dr-google-how-the-tech-giant-is-laying-claim-to-health-data-11578719700; Sidney Fussell, "The Sneaky Genius of Facebook's New Preventive Health Tool," *Atlantic*, January 8, 2020, https://www.theatlantic.com/technology/archive/2020/01/facebook-launches-new-preventative-health-tool/604567.

34 Dieter Bohn, "Amazon Announces Halo, a Fitness Band and App That Scans Your Body and Voice," *Verge*, August 27, 2020, https://www.theverge.com/2020/8/27/21402493/amazon-halo-band-health-fitness-body-scan-tone-emotion-activity-sleep.

35 Jim Thatcher, David O'Sullivan, and Dillon Mahmoudi, "Data Colonialism through Accumulation by Dispossession: New Metaphors for Daily Data," *Environment and Planning D: Society and Space* 34, no. 6 (2016): 990–1006.

36 In Shoshana Zuboff, *The Age of Surveillance Capitalism: The Fight for a Human Future at the New Frontier of Power* (New York: PublicAffairs, 2019), 224.

Agnieszka Kurant

The contemporary human is an assemblage of agencies, from microbes to AI, that operate simultaneously. While AI scientists and philosophers are debating the agency and selfhood of machines, it is important to first analyze the agency and selfhood of post-Fordist, post-digital humans. In which direction are we humans currently evolving? What are the implications of the transformations of the human for future societies?

Contemporary microbiome research has proven that bacteria and viruses are necessary actors in our immune system, and that our microbiome impacts our mental states and our decision-making processes. Simultaneously, our decision-making processes are constantly being hacked by corporations via algorithmic data mining. The contemporary human is being hacked both from the inside—by microbes—and from the outside—by algorithms, which constantly manipulate and optimize our decision-making processes. As a result, the concept of the self as autonomous individual begins to collapse.[1]

We used to justify human exceptionalism by our ability to make independent, reasonable decisions, following our feelings and judgements. Today, however, it turns out that there are situations in which algorithms can give us better counsel than our feelings about which route to take, what to read or watch next, what to study, whom to date, whom to vote for. In medical diagnostics and in engineering, decisions made by algorithms have, in some cases, proven better than those made by experts. This automation and optimization of human decision-making is an evolutionary change that might one day, in retrospect, be compared to agricultural evolution or the invention of writing.

Automated Crowds

The hacking of human decision-making affects not only individuals but the collective decision-making of entire societies. As philosopher Franco "Bifo" Berardi points out, automation not only changes the organization of work but also implies the end of conscious political decision-making, replacing conscious voluntary choices with algorithmic procedures.[2] "Cognitive automatisms," he writes, "lead to the emergence of a form of intelligent behavior that is not consciously performed by the individuals who take part in the swarm." Today, "social action is less and less the result of consciously organized choices, and more and more the result of automated chains of social interaction."[3] Automation of human decision-making means automation of the general intellect, which becomes "fragmented and dispossessed of self-perception and self-consciousness."[4]

Today, when algorithms can quantify the aggregated value of social capital precisely, humans have become the energy fueling the digital economy. We are the new oil. While the twentieth

1 Tobias Rees, *After Ethnos* (Durham, NC: Duke University Press, 2018).

2 Franco "Bifo" Berardi, "The Neuroplastic Dilemma: Consciousness and Evolution," *e-flux Journal*, no. 60 (December 2014), https://www.e-flux.com/journal/60/61034/the-neuroplastic-dilemma-consciousness-and-evolution.

3 Berardi, "The Neuroplastic Dilemma."

4 Franco "Bifo" Berardi, *The Uprising: On Poetry and Finance* (Los Angeles: Semiotext(e), 2012, 8).

5 Trebor Scholtz, ed., *Digital Labor: The Internet as Playground and Factory* (New York: Routledge, 2013).

6 "Agnieszka Kurant on Collective Intelligence," *frieze*, no. 200, January-February 2019, https://www.frieze.com/article/agnieszka-kurant-collective-intelligence-0.

7 Catherine Malabou, *What Should We Do with Our Brain?*, trans. Sebastian Rand (New York: Fordham University Press, 2008).

century was about extraction of fossil fuels, the twenty-first century is about the extraction of our data. Each one of us constitutes millions of data points. We are a giant, distributed social factory of data production and exploitation, in which we all perform invisible labor on a global conveyor belt operated by corporations and governments. We no longer know when our labor or social capital is being stealthily harvested. This concealed exploitation of the energies of the social collective intelligence, or general intellect, materializes in the phenomenon of "playbor," where value is harvested unbeknownst to the worker through labor masked as play.[5]

Corporations such as Google, Amazon, and Facebook capitalize on the dispersed and networked activity of their multitudes of users. Corporations and governments quantify our enthusiasm, joy, fear, anger, and disgust for advertising or propaganda purposes. Social energies have become part of the energy market: just as coal, oil, and gas are extracted from the natural world, social energies are being mined and precisely quantified by algorithms. The value and influence of "people power" can be calculated using social physics and quantitative analysis. The contemporary global economy is based on flows and conversions of energy into information, into capital.[6] New technologies enable increasing capacities of social mobilization and political agency. In this situation, both the individual and the crowd are evolving into new, mutated forms. The digitally mobilized crowds are, simultaneously and paradoxically, controllable and unpredictable.

Meanwhile, a new working class is being born. Platforms such as Amazon's Mechanical Turk are outsourcing various microjobs to hundreds of thousands of people working online all over the world. These digital workers receive micropayments for simple tasks. Often, these are tasks that computers cannot do, which is why Amazon's founder, Jeff Bezos, named this form of labor "artificial artificial intelligence." The structure of these kinds of outsourcing platforms is based on collective intelligence that could potentially be harnessed to solve problems for the benefit of the public good, much like Wikipedia does today. For the moment, however, the tasks posted on Amazon's platform are mostly corporate, and often exploitative. The so-called Turkers, earning on average less than one dollar an hour, become a new working class with no protections and no means of unionizing. The system has found a perfect worker.

Twenty-first-century scientists and philosophers such as Catherine Malabou study the effects of epigenetics and the neuroplasticity of the human brain.[7] But what are the consequences of the current plasticity or evolution of the social brain caused by the automation of our decision-making? What are the effects of these civilizational changes for our collective unconscious? We have become, as philosopher Yuval Noah Harari pointed out, "hackable animals."[8] Today's technology produces empirical examples of the modern concepts of collective agency, such as Karl Marx's notion of the general intellect, Gilbert Simondon's domain of the transindividual, or Antonio Negri's labor of the multitude. As Matteo Pasquinelli points out, "machine intelligence is not anthropomorphic but *sociomorphic*: it imitates and feeds on the *condividual* structures of society rather than the individual ones."[9] "Machine intelligence mirrors social intelligence," and therefore uses similar mechanisms of social control.[10]

In 2020, the Chinese social credit went live. Chinese credit scores—analogous to those computed by American credit-reference agencies, such as Experian and Equifax—look not at consumer creditworthiness but at social behavior, with the criteria for desirable behavior defined by the Communist party. The system is a result of operations performed by the so-called police cloud, in which police gather and synthesize all the information they can: medical histories; food and courier deliveries; supermarket loyalty card numbers; methods of birth control; religious affiliations; online behavior; airplane and train journeys; GPS movement coordinates and biometric data; face, voice, fingerprint, and iris scans; and DNA. The project aims to form a new kind of human being, one who has fully internalized the demands of the state and is able to self-censor, self-monitor, and self-supervise.

According to Berardi, citing theorists such as Kevin Kelly, "The social organism is going to be transformed into a swarm, a collective organism composed of many individual bodies whose brains follow rules embedded in the nervous system of each individual. The individuals who comprise a swarm act in a coordinated way even if no centralized structure of control dictates how they should behave."[11] In order to better understand these transformations of human societies, we have to abandon notions of both human and machinic intelligence as individual, and instead consider each as comprised by a multitude of agents, a collective intelligence.

Extended Mind

Individualized intelligence, and the belief that each person's independent intelligence is located in their brain, is ingrained into Western culture. However, scientists have proven that intelligence is never individual but rather distributed. Cognitive scientists and philosophers such as Andy Clark and David Chalmers describe the extension of the mind into the external world as distributed cognition, while anthropologists such as Tobias Rees suggest that fields like AI research, microbiome studies, or gene-editing powerfully undermine the distinction between the human and nature, or between humans and machines.[12] According to the advocates of extended cognition, the physical

8 Yuval Noah Harari, "The Myth of Freedom," *Guardian*, September 14, 2018, https://www.theguardian.com/books/2018/sep/14/yuval-noah-harari-the-new-threat-to-liberal-democracy.

9 Matteo Pasquinelli, "Abnormal Encephalization in the Age of Machine Learning," *e-flux Journal*, no. 75 (September 2016), https://www.e-flux.com/journal/75/67133/abnormal-encephalization-in-the-age-of-machine-learning. Emphasis in the original.

10 Pasquinelli, "Abnormal Encephalization in the Age of Machine Learning."

11 Berardi, "The Neuroplastic Dilemma."

12 Andy Clark and David J. Chalmers, "The Extended Mind," in *The Extended Mind*, ed. Richard Menary (Cambridge, MA: MIT Press, 2010), 27–42; and Rees, *After Ethnos*.

13 Simon Schaffer, "Of the Body of Politics and the Body of Nature," in A Book of the Body Politic: Connecting Biology, *Politics and Social Theory*, ed. Bruno Latour, Simon Schaffer, and Pasquale Gagliardi (Venice: Fondazione Giorgio Cini, 2020), 282.

14 Peter Godfrey-Smith, *Other Minds: The Octopus, the Sea, and the Deep Origins of Consciousness* (New York: Farrar, Straus and Giroux, 2017), 103. Emphasis mine.

15 Marvin Minsky, *The Society of Mind* (New York: Simon & Schuster, 1986).

16 Jessica Riskin, *The Restless Clock: A History of the Centuries-Long Argument over What Makes Livings Things Tick* (Chicago: University of Chicago Press, 2016), 344.

17 Daniel Dennett, *Consciousness Explained* (New York: Back Bay Books, 1991).

18 Rodney A. Brooks, "Intelligence without Reason," in *IJCAI '91: Proceedings of the 12th International Joint Conference on Artificial Intelligence*, ed. John Mylopoulos and Ray Reiter (San Francisco: Morgan Kauffman, 1991), 1:569–95.

machinery of the mind itself sometimes extends beyond the skull and skin.

Complexity science—a field studying complex systems as opposed to linear flows of cause and effect—introduced the notion of collective intelligence into both science and culture. Collective intelligence—observed in slime molds, termite colonies, the internet, social movements, the stock exchange, urban fabric, and our brains—is a phenomenon where novel forms emerge in nonlinear and unpredictable ways out of interactions between thousands or millions of elements or agents in a complex system. Groups may have minds in much the same way that individuals have minds. It has been observed that various termite colonies display disparate collective behaviors as varied as human personalities. Some colonies are collectively more dynamic, curious, adventurous, or militant, while others are more sluggish and prudent. Complexity scientists have made similar observations about other systems of collective intelligence, such as social movements, which display diverse personality traits as super organisms or "collective persons."[13]

What might it be like for a subjective experience to be *plural*—"we" and "not-we" instead of "I" and "not-I"? Peter Godfrey-Smith describes the octopus as an existing form of plural consciousness, noting that most of the octopus's neurons are not clustered in a centralized brain, as they are in vertebrates, but located in the arms, which seem to have some degree of autonomy. "For an octopus, its arms are partly *self*—they can be directed and used to manipulate things. But from the central brain's perspective, they are partly *non-self* too, partly agents of their own."[14] They form a distributed intelligence.

An anti-Cartesian conviction permeates the work of thinkers such as philosophers/cognitive scientists Andy Clark and Daniel Dennett, as well as evolutionary roboticist Rodney Brooks and writer Kevin Kelly. They all propose that there is no such thing as a unified, rational self. There is no central locus of control. There is, inside your brain, a society of different minds, writes Marvin Minsky, author of the first neural-network simulator.[15] The brain is a multitude of agencies. A person is nothing beyond various subagencies and processes in the nervous system. Thus, as Jessica Riskin notes, "Descartes's rational self is banished from current artificial intelligence and cognitive science as well as from behavioral robotics."[16] The very smallest, lowest-level agents of thought are unintelligent and indeed cannot think at all. As Dennett points out, agency cannot inhabit the acting creature but can only be ascribed to it from the outside.[17]

Emergence

At the heart of contemporary systems sciences, from living systems and geology to economics and AI, is the notion of emergence: an agencyless mode of coming-to-be.[18] Emergence is an organizing idea that has permeated the natural and social sciences during the last three decades. Among the paradigmatic emergent phenomena is self-organization, which constitutes a form of "minimal biological agency" attributable, for example, to cells. This agency resides not in any of the agents but in their organization. Emergence denotes any complex phenomenon that seems irreducible to the sum of its parts: phase transitions, as from solid to liquid or liquid to gas; hurricanes; living organisms; the stock market.

According to Rodney Brooks, thought and consciousness will not need to be programmed in machines; they will emerge. Brooks and other researchers in the field of evolutionary robotics work to evolve AI as an emergent result of assemblages of behavior-generating modules.[19] The crucial feature of these robotic creatures is that they have no Cartesian, unified self, no central governing module, no central control system. It turns out that intelligence can exist without a unified, knowing subject. In these creatures, as in nature, intelligence is in the eye of the observer.

Analyzing these phenomena and concepts, we start to realize that once the notion of the autonomous, integrated self starts to collapse, we also need to question the notion of individual authorship. Artistic creativity and value production in art undergo a mystification that operates via one of the most common myths surrounding art: the idea of creativity as an individual process.

The concept of the "author" is, to an important extent, a construct created to appropriate the labor of the multitude in order to proclaim it as an individual's artwork. Mechanisms in the art market and cultural industries lead to the exploitation of the creative general intellect. They are based on the fetishization and speculation of not only cultural and artistic objects but also creative processes and ideas. The very figure of an author represents a mechanism devised to appropriate value produced in a dispersed and networked creative process.[20] In the future, individual authorship might be replaced by more complex, collective forms.

Artificial Societies

Many of my artworks consider complexity science, a branch of science most significantly established at New Mexico's Santa Fe Institute, which focuses on growing artificial societies. The agent-based models developed there have allowed for the testing of social theories on a large scale. Artificial societies are based on the process of emergence from the lower microlevel of a social system to a higher macrolevel. Since the late 1960s, economists and social scientists have used computer simulations of societies or groups of people to analyze socioeconomic phenomena and observe agent-based complex system behavior. Craig Reynolds (who continued the earlier experiments of John von Neumann, John Conway,

19 Brooks, "Intelligence without Reason."

20 Agneiszka Kurant, Jan Sowa, and Jakub Szreder, "Do We Need a Lab? Capture, Exploitation and Resistance in Contemporary Creative Communities," afterword to *Joy Forever: The Political Economy of Social Creativity*, ed. Kurant et al. (London: MayFly, 2014), 249–58.

21 Stefan Helmreich, *Silicon Second Nature: Culturing Artificial Life in a Digital World* (Berkeley: University of California Press, 1998).

22 Joshua M. Epstein and Robert Axtell, *Growing Artificial Societies: Social Science from the Bottom Up*, (Washington, DC: Brookings Institution; Cambridge, MA: MIT Press, 1996).

23 Jonathan Rauch, "Seeing Around Corners," *Atlantic*, April 2002, https://www.theatlantic.com/magazine/archive/2002/04/seeing-around-corners/302471.

and John Holland) attempted to model living biological agents, a method known as artificial life.[21] Other scientists, such as Thomas Schelling, experimented in the 1970s with complex models of evolutionary game theory—a strategic interaction among rational decision-makers, initiated by von Neumann in the 1920s and developed by the RAND Corporation in the 1950s as complex models of the prisoner's dilemma. Later on, computational methods of artificial life were applied to the analysis of social systems, called "artificial societies" by epidemiologist Joshua M. Epstein and computer scientist Robert L. Axtell. Artificial societies use interactions between "agents"—these could be virtual people, groups, machines, or animals—to simulate a complex system, test social theories, or predict future social phenomena.

In the 1990s, Epstein and Axtell decided to use emergent processes seen in coral reefs and flocks of birds to grow a simple artificial world in which small hunter-gatherer creatures would move around a landscape, finding, storing, and consuming the only resource—sugar.[22] As one reporter put it, "When they brought what they called Sugarscape to life with a computer, they were startled to see that, almost immediately, their rudimentary artificial society produced a skewed distribution of sugar that looked very much like the skewed distribution of wealth in human societies, even though nothing about the agents' simple behavioral rules pointed to any such outcome. For several years, Epstein and Axtell developed and elaborated Sugarscape, and discovered that simple rules could produce complex social phenomena that mimicked migrations, epidemics, trade. They began applying their technique, which they called 'agent-based modeling,' to a variety of problems and questions."[23] Today, complexity scientists apply artificial society models to grow cybercrime waves and simulate racial segregation in artificial neighborhoods, epidemics caused by bioterrorism, and artificial genocides. They also apply these models toward the resolution of intractable conflicts.

Artificial societies provide a new method for sociological analysis in the form of computational sociology. But until recently, these models were mere simulations and proved to be inaccurate in comparison with real human societies. Now, however, we can use neural networks and real-life data input to create more realistic models of societies and test certain social or economic theories and solutions for a better understanding of the transformations of the human.

The observation of collective intelligence in both nature and culture allows us to imagine a creative AI emerging from a collective intelligence of humans and machines. Artificial intelligence researchers often ask if artistic creativity is computable and whether AI can produce artworks. Perhaps the artificial intelligence artist should be imagined as an emergent complex system of "artificial artificial intelligence." Using the collective

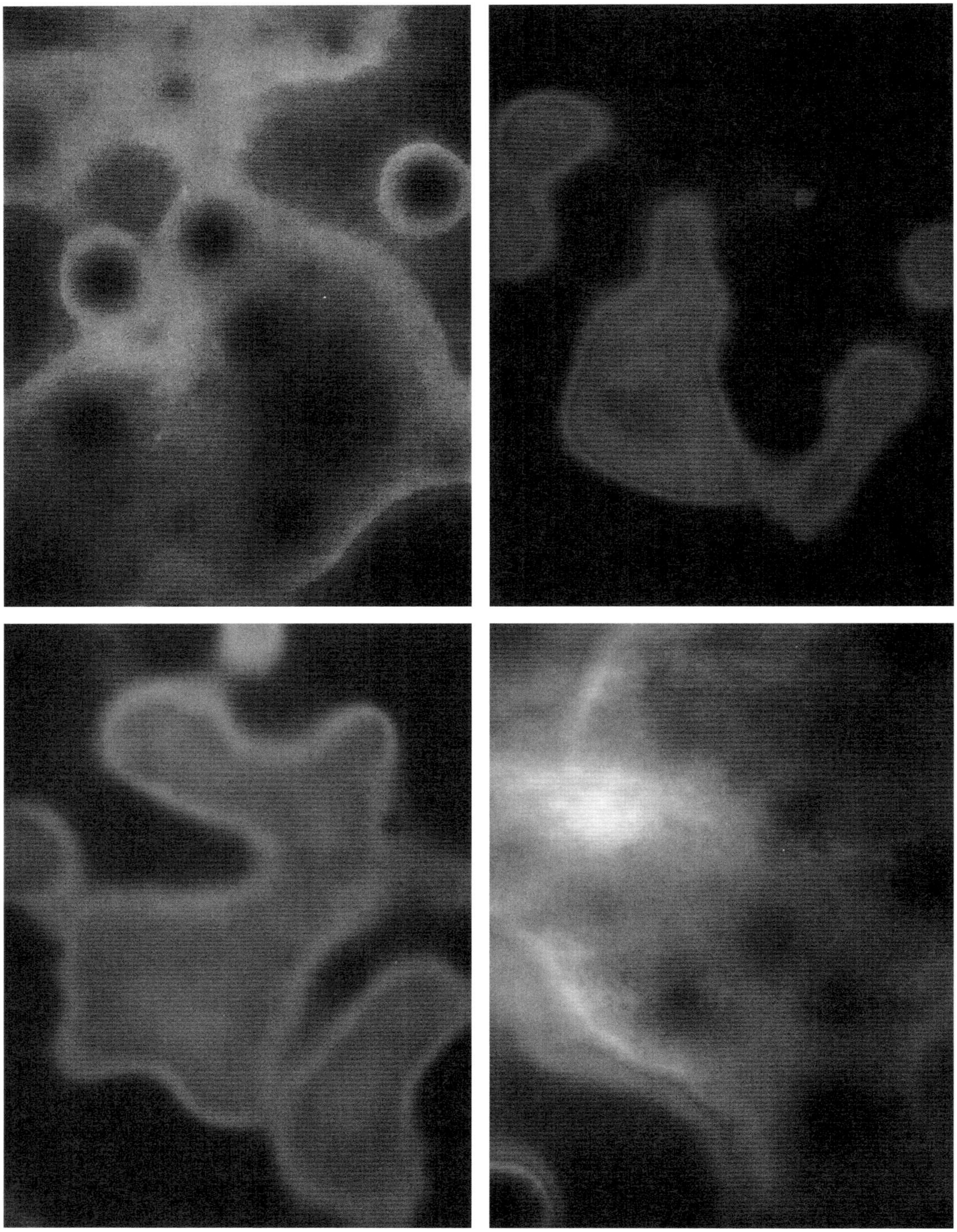

Agnieszka Kurant, *Conversions #2*, 2020. Liquid crystal ink on copper plate, Peltier elements, Arduino, custom programming, transistors. Engineering: Agnes Cameron, MIT; Nick Wallace

intelligence of crowdsourcing to simulate the nonlinear, complex system of the billions of neurons in a human brain is currently the closest we can get to simulating the emergence of a single thought or of consciousness in general.[24]

Drawing on these natural and social phenomena, I started to employ an alternative way of producing artworks by applying collective intelligence to art production. This has led me to grow or evolve artworks like living organisms or geological formations, crowdsourced to the collective intelligence of thousands of nonhuman and human agents, from molecules and bacteria to online workers. My work centers on general morphogenesis—on the birth of forms and the ways in which they are generated as outcomes of collective intelligence, whether a termite mound, a mineral, a plant, a sign, an image, a tool, an animal, a crowd, a meme, or a revolution.[25]

A.A.I. (2017) employs termite colonies; *Assembly Line* (2016), a work based on profit sharing, involves thousands of people—the new digital working class; *The End of Signature* (2013–ongoing) relies on communities of people supporting a common cause, such as workers unions or museum visitors; *Animal Internet* (2017) relies on robotic swarms; and *Conversions* (2019) uses the behavioral data and emotions of members of protest movements.

Many of my works emulate nature and behave like living organisms or self-organized complex systems in order to reflect the interplay of human and nonhuman agencies in the world and the ways in which individual authorship is often replaced by complex collective forms. I collaborate with computer scientists, biologists, sociologists, and anthropologists to create artworks that are alternately natural and artificial; real and synthetic; living and not; biological, geological, and algorithmic. I explore the transformations of the human and the future of labor and creativity.

Conversions is a series of works which physically react to changes happening in society.

These ever-morphing liquid crystal paintings transform according to the algorithmic *sentiment analysis* of emotions harvested from thousands of Twitter accounts belonging to the members of protest movements around the globe. Each painting uses artificial intelligence algorithms to analyze the enthusiasm, fear, anger, disgust, sadness, surprise, or joy expressed in thousands of posts. The dynamics of this collective data is then fed to an *artificial society* model—a computer simulation of social interactions.

That artificial society simulation is connected to a custom circuit board that heats or cools layers of liquid crystal pigments on top of a copper plate. The shifts in the artificial society influence the transformations of the liquid crystal ink. This produces a digitally and physically evolving collective thermal footprint of a multitude of people.

The work uses AI neural networks to harvest the social energy from thousands of people around the world and convert it to thermal and electrical energy. The decisions and feelings expressed by

24 Agnieszka Kurant, "Uncomputables," *Cabinet*, no. 61, Spring-Summer 2016.

25 Nicolas Bourriaud, *Inclusions: Aesthetics of the Capitalocene*, trans. Denyse Beaulieu (London: Sternberg Press, 2021).

individuals all over the world cause the paintings to evolve, like an organism or ecosystem. These collectively and unwittingly authored paintings are portraits of the collective intelligence of aggregated social capital, a perpetual litmus test of social energies—the evolving collective footprint-signature of the multitude of citizens.

The series of sculptures *A.A.I.* is the product of the collective intelligence of termites to whom I outsourced my art production. This method was inspired by the outsourcing of industrial production to Asia and Africa and the exploitations present in the mining industries. Termites, just like humans, are one of the few species in nature that have evolved to form complex worker societies; the by-products of their labor are mounds with architectural qualities, often resembling the ruins of civilizations, pyramids, or cathedrals. Working with entomologists in laboratories at the University of Florida, I employed an entirely unaware worker society of millions of termite specimens to produce sculptural mounds. I supplied the termites with alternative building materials, such as vividly colored sands, crystals, and gold, resulting in the creation of hybrid forms hovering between nature and culture. The project is a model of a dispersed social factory and of the concealed exploitations of social energies present in a contemporary economy, where companies such as Google, Amazon, and Facebook are harvesting our personal and behavioral data and selling it to advertising companies.

It has been observed that various termite colonies display disparate collective behaviors resembling various human personalities. With this work, I explore how communities of people might also be perceived as collective organisms with distinct personality traits. Entomologists are able to describe the "collective personality" of a given termite colony by looking at the shape of a given termite mound. *A.A.I.* proposes an alternative way of producing artworks—through crowdsourcing. Singular authorship is slowly being replaced by complex collective forms, as in the case of Wikipedia, where knowledge is produced collectively. These hybrid sculptures are the result of a collective effort of entire animal societies, and reference contemporary memes circulating in culture.

The point of departure for *Post-Fordite* (2019–20) is a recently identified quasi-geological formation—Fordite, also known as Detroit agate, formed through the quasi-fossilization of countless thin layers of both mechanically and hand-sprayed automotive paint on the tracks, pallets, and floors of car factories in cities such as Detroit, built up over more than 100 years of factory operation. Since the factories went out of business and closed, unemployed workers and various scavengers have been penetrating the abandoned production lines for valuable leftovers. Since Fordite can be cut and polished, fragments of fossilized Fordite started circulating online, accruing value and treated as precious stones to produce jewelry. I buy fragments of fossilized

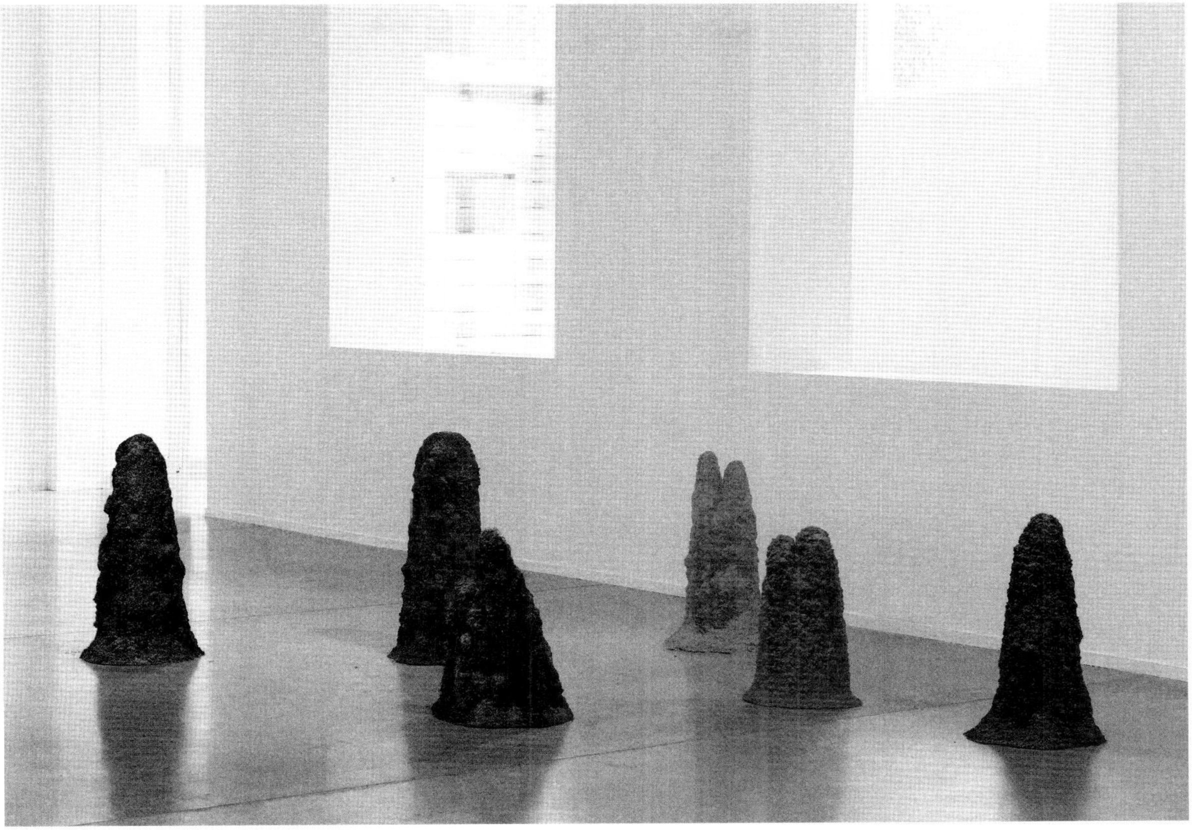

Agnieszka Kurant, *A.A.I. (Artificial Artificial Intelligence)*, 2017. Termite mounds built by colonies of living termites out of colored sand, gold, and crystals. Collaboration with Dr. Paul Bardunias, La Panacée, Montpelier

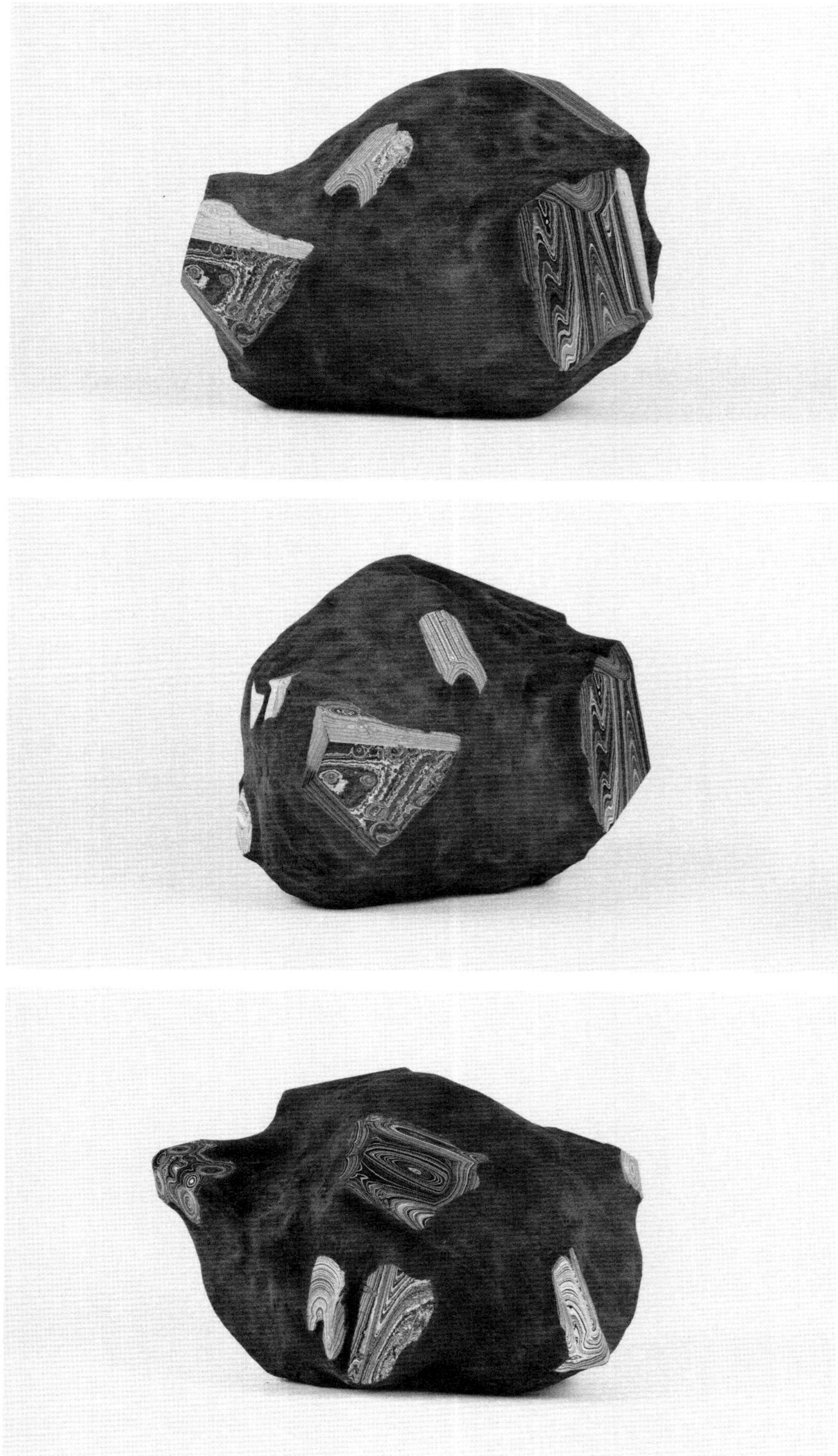

Agnieszka Kurant, *Post-Fordite 5* and *Post Fordite 6*, 2020. Fossilized automotive paint, resin, foam, steel, acrylic paint

Fordite found at various factories around the globe and combine them together into one abstract speculative rock formation, anticipating the geological changes caused by post-Fordism. *Post-Fordite* is an aggregation and amalgamation of the social capital and labor of generations of workers of the now automated industry.[26]

Shifting fashions and fads in car colors and finishes over the past 100 years can be traced in the quasi-geological strata of the fossilized enamel paint.

Each facet of *Post-Fordite* reveals a different interior, and what emerges is an n-dimensional, impossible space evoking M.C. Escher's drawings, as if inside this speculative rock, various geologies existed at the same time.

Post-Fordite imagines a geological formation of post-Fordism—the transformation from factory labor into immaterial labor, during which factory production is largely replaced by immaterial products such as ideas, copyrights, patents, automation, and information technologies, and everybody becomes a worker on the global conveyor belt of data production and exploitation. Joseph Beuys's idea of everyone being an artist can be converted into "everyone is a worker," as we all leave our digital traces in the virtual geology through our creative actions performed online, constantly mined by corporations.

In *Aggregated Ghost* (2020) and *Assembly Line*, I hired ten thousand online workers from the Amazon Mechanical Turk crowdsourcing platform to submit their self-portraits. Using neutral network artificial intelligence algorithms, which I developed with computer scientists at MIT Artificial Intelligence Lab, I fused all these images into a collective self-portrait of this new online working class. The profits from the sales of these works are redistributed among the workers who participated.

Similarly, *Production Line*, developed jointly with the artist and writer John Menick, amalgamates the labor of thousands of workers from online crowdsourcing platforms to virtually produce a global production line, here materializing as a physical composite line, fragments of which are drawn by thousands of people in various parts of the globe. *Production Line* is a physical embodiment of the entire society becoming a gigantic, dispersed social factory.

In the series *The End of Signature*, I analyze communities, social movements, and societies as super-organisms or collective persons with personality traits, and I develop collective signatures for them. To create *The End of Signature*, I worked with computer programmers at MIT to develop software that morphs a group of signatures into one collective signature. The project took various forms. For example, at the Guggenheim Museum, visitors were invited to contribute their own signatures, which were subsequently scanned and aggregated into a single form that was projected onto the exterior of the iconic Frank Gehry building for the duration of the exhibition. An Autopen automaton machine also formed part of the work. Autopen is a device used for the automatic signing

26 Diederich Diederichsen and Oier Etxeberria, eds., *Cybernetics of the Poor* (Berlin: Sternberg Press, 2021).

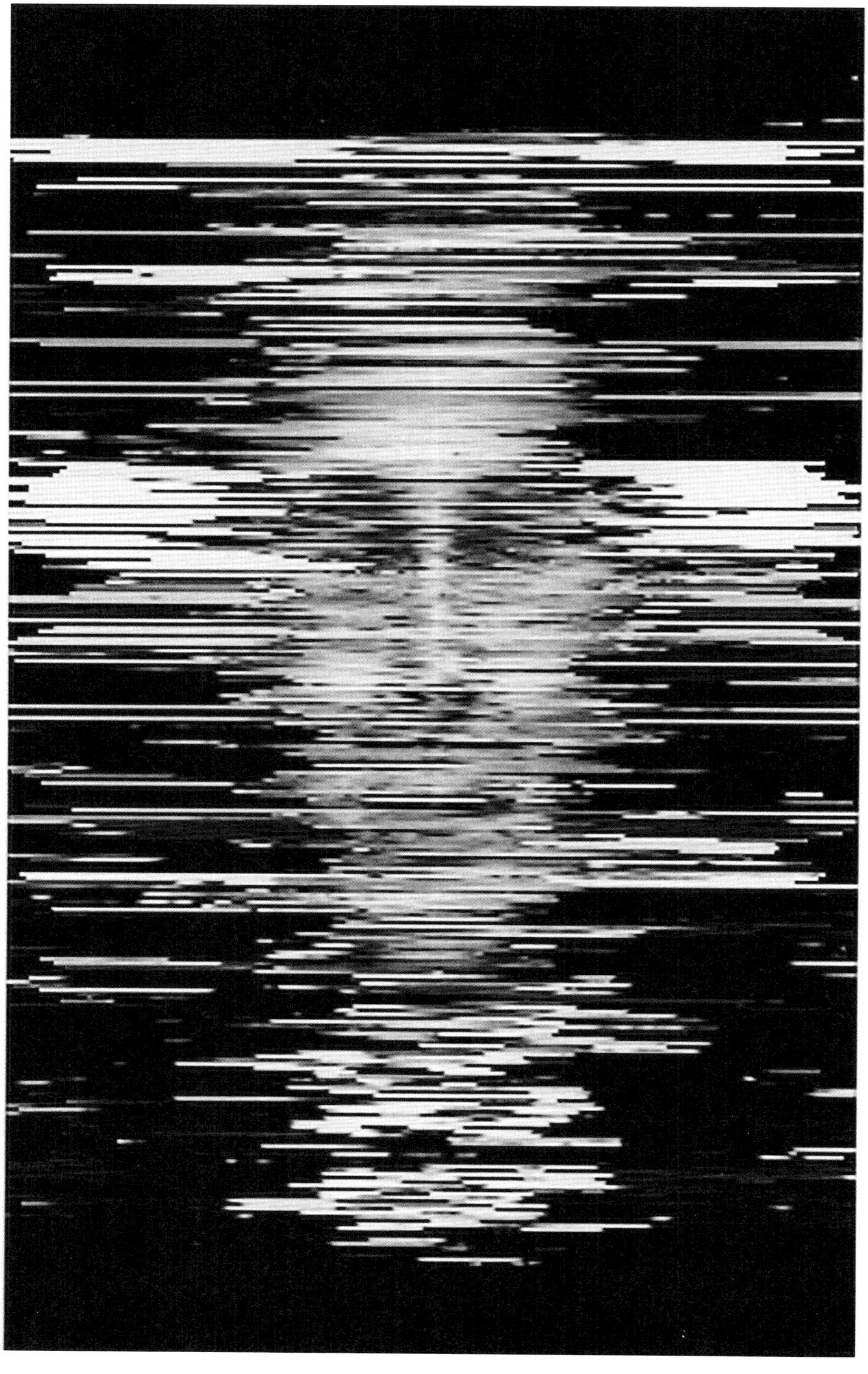

Agnieszka Kurant, *Aggregated Ghost*, 2020. Backprinted ink on acrylic mounted on dibond. Realized with the participation of the workers of the Amazon Mechanical Turk online crowdsourcing platform. Produced by MIT CAST, in collaboration with prof. Boris Katz, Andrei Barbu and David Mayo

of a signature. It allows for endless duplicates of a "handwritten" signature of a given person. In my work, the arm of the Autopen machine perpetually signed fully identical collective signatures on blank sheets of paper, representing a headless authority. The work also addresses the obfuscation of individual authorship in our civilization, as well as the decline of handwriting and its gradual replacement by keystrokes on technological devices. In the near future, handwriting may become obsolete, and the notion of authorship will become even less clear with the proliferation of platforms such as Wikipedia and other forms of crowdsourcing. The piece was also executed on the façade of the Cleveland Museum of Art (as a collective signature of all the employees from the guards and cleaners to the curators and the board) as well as in public space in Utrecht (on top of a housing project), and as an indoor installation with circulating black ink and water. The newest *End of Signature* works are custom engineered LED light signs operated by a computer program, highlighting the constant circulation of social energies through a comparison to electrical energy. One of the upcoming iterations will fuse together the social capital of all the scientists and academics who ever worked at MIT. This collective signature will be a permanent installation launched in 2021 at the MIT List Center and Kendall Square in Cambridge, MA.

Collective Rorschach Test (2019) is a lenticular print featuring an evolution of memes from an online social experiment that took place on Reddit in 2017. The experiment, called Place, demonstrated a collective intelligence. More than one million people from around the world collectively built flags, fictional characters, portraits of politicians, and forms such as Black Void, a black amorphous form resembling a virus, a slime mold, or a Rorschach inkblot.

Animal Internet was realized in collaboration with a team of scientists and engineers at MIT. This project draws on the artificial society experiments of the Santa Fe Institute and the phenomenon described by scientists as "animal internet," wherein hundreds of thousands of people experience nature through their computer screens. Some fifty thousand creatures around the globe—including whales, leopards, flamingoes, bats, and snails—are equipped with digital tracking devices. Webcams installed by scientists in jungles, plains, and savannas capture the behavior of completely unaware wild animals. These animals have parallel lives—one in their natural habitat and one in the human network, in which they have names, Facebook pages, and thousands of followers.[27] This produces a hybrid, augmented reality. Drawing upon these phenomena, I imagined a collective Tamagotchi—a form hovering between nature and culture. Tamagotchis, developed in Japan in 1996, are egg-shaped, handheld digital toys that simulate live organisms. They must be regularly nourished, pet, and walked. There have been seventy-six million Tamagotchis sold to date—a

27 Alexander Pschera, *Animal Internet: Nature and the Digital Revolution* (New York: New Vessel, 2016).

nation's worth. I imagined what would happen if all the Tamagotchi owners together took care of just one single Tamagotchi, an artificial collective organism, much like a slime mold. Social movement could be rendered as the physical movement and behavior of this hybrid creature. *Animal Internet* presents a group of artificial animals operated by the decisions and actions of thousands of people around the world. The work is based on animatronic mechanisms covered with fur to emulate mammalian life. These artificial animals are placed in terrariums with artificial landscapes that include real plants and rocks, and they are available to watch via webcam. The piece juxtaposes two real webcam broadcasts—one of a wild tiger in the jungle and one of a polar bear in the Arctic, both filmed unbeknownst to them—with two webcam videos of the artificial animals. The artificial animals' movements are operated by crowdsourced online workers around the world, who are tasked with expressing in writing how they are feeling at a given moment. Their answers are aggregated into one overall behavior, which shifts depending on the collective state of mind of the prevailing majority of workers. This result is translated into the movement of the fictional furry animal observed in the webcam and into a collective behavior of an artificial gerbil swarm.

As the notion of an integrated, individual "self" is no longer tenable, my works question the concept of artistic creativity as an individual process. In the future, individual authorship might be gradually replaced by hybrid, complex, collective forms integrating crowdsourcing, AI, microbes, viruses, and other nonhuman agencies. In this context, my work proposes a novel way of producing artistic forms based on the crowdsourcing, aggregation, fusion, and amalgamation of inputs from thousands of human and nonhuman agents. I explore the ways in which collective intelligence and nonhuman intelligence (animal, microbial, and AI) undermine our understanding of nature and culture. I try to draw attention to the necessity of abandoning the notion of the human as an individual intelligence and of rethinking the human as an assemblage or polyphony of agencies and intelligences, from microbial and viral to artificial. Drawing on artificial life, self-organizing matter, transformations of energy, and complex systems, I fuse collective intelligence and artificial intelligence to investigate in which direction humanity and nature are currently evolving, as well as how to amend and self-organize the general intellect, or the multitude.

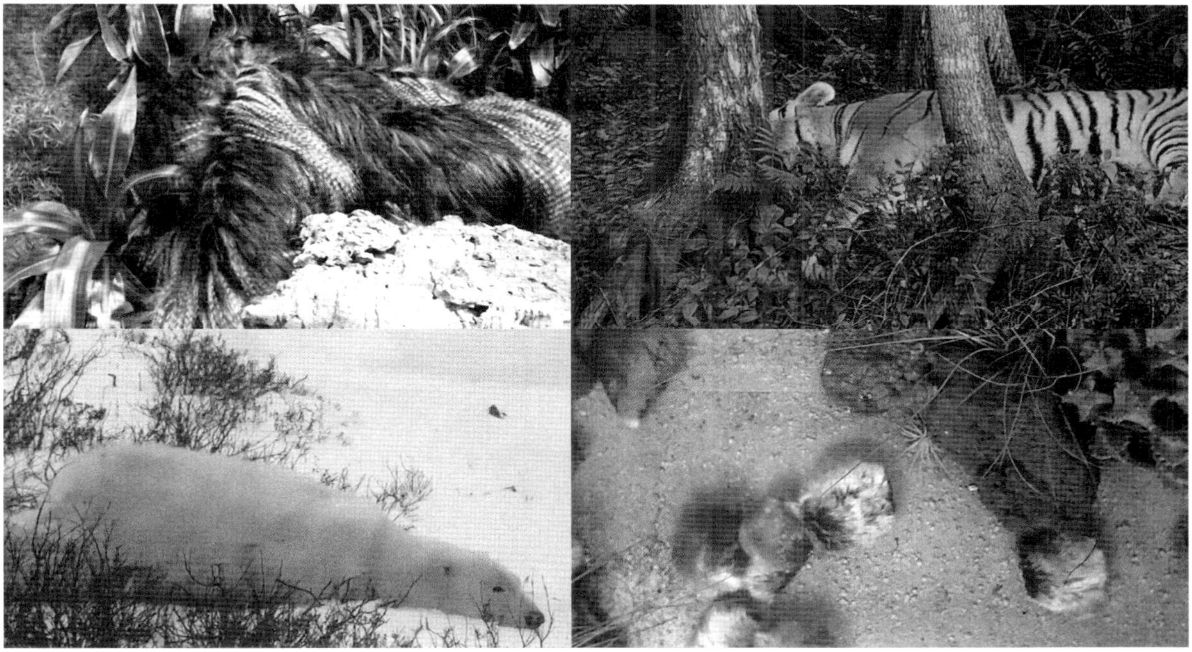

Agnieszka Kurant, *Animal Internet*, 2018. Dual-channel video installation. Engineering and collaboration: Adam Haar Horowitz, Owen Trueblood, Ishaan Grover, Agnes Cameron, and Gary Zhexi Zhang

LOU CANTOR: We talk a lot about how our practices reflect the global situation and about how this conversation will be shaped by factors way beyond our understanding—not to mention our control. The pandemic sent us on an unplanned trip, and the small selection of English-speaking TV channels in the hotel we stayed in informed us about the monsoon flood you are experiencing in Jakarta. I'm happy you were able to call . . .

FARID RAKUN (OF RUANGRUPA): It's not as bad as they say in the press, or I shall say it is always bad but it's not shocking, because we have it almost every year.

LC: I must say, it's hard to imagine that something of this scale is a recurring problem. Where I come from [Poland], we have little things—for example, when it starts snowing, city services are a bit late with cleaning streets, and year after year the evening news repeats the same sentence: "Winter surprised road workers." We tend to dramatize it, but it's just an inconvenience.

FR: Maybe you've heard this, but the reason why it's happening is that Jakarta is sinking. So it's not surprising that this will get worse and worse every year. There are measures taken to counteract it, but whether they are the right measures is a subject of debate. If you asked me, I'd say we have to accept what we've done and try to move on rather than fight nature. It's too little too late. Floods have been occurring since the kingdom times, so even precolonial, let's say. Some years are worse: 2007 for example, or 1991. There was a particularly bad year during my childhood. But now, the problem is being layered, and we're fighting a losing battle. The way to go is to live with the water, if you ask me. But then when it becomes a practical choice for the governance . . . The political discussion got too complicated, so I don't know.

LC: The way you describe it makes me feel that this reoccurrence or repetition of interlocking periods of aggravation and mitigation progresses simultaneously in every direction and occupies all aspects of the space we inhabit. It's just easier to lose sight of it when you are not confronted with this parabola of a monsoon. At the same time, it might be simply impossible to solve the problem. There are so many contradicting interests and plenty of valuable things that have to be weighed against each other. If you know beforehand that some will have to be lost, the only answer that respects everyone and also the values you believe in might make you restrained from making any decision at all. We are sometimes glad that we work in culture, on a smaller scale and incomparably lower complexity than governance, and yet I feel we share an observation that our institutions resort to mitigation: they address the problems such that the grammar of the question excludes the possibility of an answer.

1 For documenta fifteen, Ruangrupa has extended invitations to a number of artistic initiatives from different contexts in the world. If a lumbung (directly translatable as "rice barn" or "granary") is understood as a network, these initiatives are the members, with whom ruangrupa co-imagines, co-develops and co-creates documenta, while co-governing the resources that render such even possible.

FR: Yes, we talked about this before: I think the problem is in the bones. Right now, many institutions are trying to change themselves, but those changes are only skin-deep. It's not only happening in Europe, but also in China, South Korea or Japan, Singapore, and even here in Indonesia. If those institutions want to address the problems that they're working on right now, they need to change the bones as well. It's as if their mandate is also a source of the problem. Our perspective is that it's a matter of scale. If you want to reach the globe from the get-go, you need a certain scale, and the scale is a problem. That's why we trust small scales all the time: it's not a magic pill, but we need to be small to keep growing.

LC: This is something that keeps coming up in our conversations too, isn't it? Yuval Noah Harari has this theory, which I find very compelling, that claims humans as a species are particularly good at collaborating. Yes, other animals can do it, but we just happen to be more dependent on it. For every species, there is a specific number of individuals who can have intimate knowledge of each other. And that, of course, does not have to do with sexuality or desire but with a form of trust: that indescribable feeling that this other person is on your team, that they're not going to leave you behind. "Dunbar's number," which measures humans' cognitive limit, is 150 individuals. I was thinking about this when you mentioned the necessity for being small. I trust that intuition. I also think that once you try to be global, you are necessarily losing a lot, that there are many things that you cannot do on a larger scale. This is a major problem that the institutions you're talking about have: they don't realize that the scaling-up becomes an empty ritual, instead of building anything meaningful.

You mentioned "networked localities," and I wonder how you would go about implementing them.

FR: I don't think I have a practical answer when it comes to real politics. Unfortunately, our work is still very, very limited to contemporary art, which is a small section of the world. But we're always trying to break things down when they become too big. Today, we experienced one example, and it's actually not easy: There is an assembly of documenta fifteen artists and lumbung members.[1] Right now we have fourteen members and twelve artists, but some of them are groups, so there are a lot of people in different time zones. We broke the group down into two parts, because of time zones mainly, but then today, we decided that Ruangrupa should not be present, because otherwise, we form too big of a group. I decided that I don't need to be present: I can watch the recordings, or ask questions afterwards. Many people like being invisible on Zoom and prefer to just listen. But it's hard: a lot of times you want to take part, but I realize the importance of letting go. Creating a big group is not always necessary or productive. We think of documenta fifteen, and other things that we are doing here in Indonesia, as a wide-open format, but not one that swallows everything else. How can we be generous on the level of funds and other resources, such as the attention economy or time? How can we delegate certain decisions outside of our control? It's easier in Indonesia where we speak the same language and come from the same cultural background. documenta has been much more difficult, because the tradition is for everyone to place all responsibilities on us. It's a learning process as well: if we let go, we celebrate multiplicities. It might be confusing for a lot of people, but if we give it more time and be much more generous, maybe it can be more human. Maybe we get an event that makes much more sense to experience.

LC: First, what you're saying resonates with what we think and believe in, and second, we also have had the experience that when you present this more open-ended opinion, you do run into a wall: there is some misalignment with the understanding of responsibility. I wrote to you earlier that this is something I struggle with: I fail miserably at finding words to express our position. When you are engaged in collective work, you think a lot about social responsibility: it happens by nature of working with other people. By default you have to be attentive and careful and notice the person next to you. At the same time, our reality is constructed in a manner that ignores these dynamics and favors individual responsibility backed by an ID card or other forms of legal status. This may look like just a formality, but I find it disturbing that the credibility of a group should get tied to traditional forms of identity, which can be seen as a form of policing.

FR: Hopefully, people can be as generous in their understanding as you are. Not many people are—that's because they don't have time and so they're stingy about it. Understanding of time is also subjective: different for urban species like us and those in rural cultures. To talk and work across those divisions, you need to work with a different understanding of time. Slow doesn't mean lazy—that's important. Colonial powers, throughout history, understood slow as lazy, when it's actually generous and inclusive: we need to accept other scales of time. There are two scales: size and time. Both are different for everyone. Like in *Flatland*, we are just like them when we are talking about quantum physics . . .[2] We perceive three dimensions, or you can say four dimensions with time, and how about the fifth? I think our senses are just not equipped to do that. The great thing is to actually accept that and then try to push our understanding of it. Is our understanding really so much greater than that of a frog who lives in a pond? We cannot know what's happening in Mali, or even in Denmark; we just have to trust the knowledge of our partners we are working with us there because it's a parallel world:

2 The novella *Flatland: A Romance of Many Dimensions* (1884) by Edwin A. Abbott is set in a two-dimensional world (the eponymous Flatland), in which the characters are shapes, such as circles and squares. After the protagonist, a Square, is visited by a Sphere in Flatland (who there appears as only a growing and shrinking circle), he is forced to try and conceptualize the third dimension even though it is entirely outside of his perception. He then does visit the three-dimensional Spaceland, but ironically is unable to convince the Sphere that a four-dimensional world may exist.

especially now, when we cannot travel. In a way, that makes it easier for us to let go.

LC: I guess we have to stay with the trouble . . . It's not only that it's impossible to know everything, but at a certain point, it's also necessary to accept and move on within the parameters. At the same time, we need to watch out not to mistake the echo of our own voice for communication. I think it helps to understand how many mechanisms shape our understanding of the world and how simple some of them are—how hackable if you like. Smart devices, for example, make themselves addictive with the use of the simplest tricks, the same ones you'd use to train a pigeon to play ping pong. The machines that we created to save time ended up consuming it, and they created the illusion that we're communicating in the process.

FR: Intersubjectivity and communication . . . It's a slow, crude process that requires a lot of practical things, and even if you try hard, it's still very difficult. Bringing in other voices is very hard, and we fail often. Multiplicities are really difficult to practice actually because they consist of many sidelines and are hardly ever streamlined. So many attempts fail to take off.

LC: Allowing another voice to come through requires a much greater effort than we usually think it will. Often, we are actually looking for confirmation only—this need is harvested by social media, which replaces whatever is difficult with the easiness of the confirmation bias. I think collectivity is an exercise in moving away from this. It requires listening, it requires accepting, not to look for confirmation, but rather for failure. It has to fail sometimes. Maybe our culture is wrong in calling the lack of an outcome a failure: maybe that's also just a function of communication.

FR: Right, often, you are expected to make promises and then deliver, even when the time is running out. This means you are constantly addressing promises and not the problems. I think that it also has to do with the limitations of being human—our understanding of time is constructed proportionally to our lifetime. It's actually damaging the ways we conduct things from decision-making or governance to other things beyond us: whether it's nature, a garden, curating, editing. Often, like you said, we make echo chambers. A curator making an exhibition, book, or project wants to corral forces that amplify each other and their perspectives. And I think if you understand that—and I'm following your thoughts, your projects—an algorithm could perform better than humans. We can leave that job to machines and perhaps focus our creation and editing, editorial work in different directions. We might have to change.

LC: I feel that the creation of confirmation bias is not exhibiting the weakness of technology, but of the people. One could try to create technology which does the opposite. I fear we are underestimating the power of those tools, which get ever so slightly closer to a very intimate realm in which we are constructing our beliefs, emotions, and feelings. I'm wondering how resilient older technologies—for example, democracy—will be against these forces. For political reasons, the people need a deeper understanding of the mechanism so that more technology can contribute to the progress for all and less to the benefit of the few.

FR: We've been talking to those who are keeping the old promise of the internet alive. For me, when the open-source movement started, I was really excited about what the internet could be. This goes back to our previous discussion about scale, and how to be faithful to the ideas you believed in at the beginning: Can you, if the scale grows out of proportion? The personality of the founders is also important, as it will be amplified. This monolithic model—Amazon, Facebook, Google—is not the only way to do things in the start-up economy. We could see movements like the Arab Spring, even if it failed in the long run . . . but maybe I shouldn't say it failed, since it gave birth to other things. That's one major realization that we had: we don't have to live forever, we don't have to leave heritage or legacy, we don't have to be the Rolling Stones. Thinking of technology and scale, I wanted to ask about the historic fear of fascism in cultures around you. Has it become something irrational—and what does rational thinking have to do with anything?

LC: To understand how fascism works we borrow a lot from Timothy Snyder—we are really struck by his idea of the politics of inevitability and eternity. As people we don't live reality; instead, we live narratives. In his book *Road to Unfreedom* (2018), Snyder writes that we don't consider history in politics: we do not understand it, nor do we really care for facts. Instead, in daily politics, we use narratives.

The Polish narrative, unlike the German one, is very much what Snyder calls a politics of eternity. It's built on the idea that all wrong is inflicted from outside and that time is cyclically marked by injustice coming from outside; it's also about suffering and being the Christ of Nations. By no means is this idea original or unique: for example, the fascist philosopher that Putin is following, Ivan Ilyin, builds on a similar idea of a pure nation of good and innocent people on whom evil was cast from abroad.[3] It sounds like a joke if you put it so simply, but then you have to remember that you are outside of Russia and the parameter of that narrative. But when you live inside of it, it spills over every aspect of life, beyond political or historical thought. This idea of victimhood and this weird belief that there is some form of dignity in suffering . . . It's completely absurd, but it took me a couple of years of living abroad to begin to see it. Recently, in Poland, the

3 Timothy Snyder writes about Putin's inspiration from Ilyin in "God Is a Russian," *New York Review of Books*, April 5, 2018.

4 In February 2021, professors Jan Grabowski and Barbara Engelking were ordered by a Polish court to apologize to 81-year-old Filomena Leszczyńska for writing that her uncle had given up Jews to Nazi Germans. See Jon Henley, "Fears for Polish Holocaust Research as Historians Ordered to Apologise," *Guardian*, February 9, 2021, https://www.theguardian.com/world/2021/feb/09/fears-polish-holocaust-research-historians-ordered-apologise.

5 Timothy Snyder, "The American Abyss," *New York Times Magazine*, January 9, 2021, https://www.nytimes.com/2021/01/09/magazine/trump-coup.html.

answer to what I guess is a global identity crisis is an even firmer grip on that narrative, and there were events in which some academics were questioned by the court about their publication concerned with Polish guilt in the Holocaust.[4] I don't think the court is the right organ to check if your scholarly work is sound . . . I also think the closer you stay to your narrative the closer you get to fascism . . . In his article about the mob storming the Capitol published in the *New York Times Magazine,* Snyder writes about how Trump followed strategies and rhetorics from 1930s Germany.[5] There is a striking similarity between "fake news" and Lügenpresse of 1930s Germany. Another powerful parallel from Snyder's article concerns the usage of radio by Nazis and of Twitter by Donald Trump: they both exploited the crisis of traditional press and traditional authority, the result of the great depression and of the 2008 financial crisis, respectively. Both led to many, many publishers going bankrupt, which impacted the quality of journalism. The similarities are simply too creepy to ignore. If you define fascism as this idea that a given nation is superior in any aspect, you'll start to see it everywhere—sometimes it's just concealed. Germany is my place of choice. I live here because I think it responds to the populism crisis better than other countries I know, but even here, you see dangerous narratives spilling. If Poles feel superior in their suffering, Germans think of themselves as superior in caring for nations of less capable people and even talk of "positive racism" without noticing that it is an oxymoron. This is naturally invisible from the inside, and it expresses itself when Germans see their culture not as "culture" but as the norm.

FR: Thank you for that extensive answer. I'm asking that because, as you know, we came to work with the question of locality, which so easily becomes hyperlocal. That's a trap of fragmentation. Like an Olympics composed of competing representations, not unseen in arts—just think of Venice. This model is always a competitive one, and it's not a good thing. Especially now, since the crisis has made it worse, people don't feel sufficient anymore and suddenly, they stop being generous. It does not necessarily have to do with wealth: it's about a subjective feeling of being sufficient or not. Another thing that we've learned is about narrative: everything is some form of it, including the economy, capitalism . . . just like you said. All kinds of things we see as a source of problems are narratives, so we should learn how to build counternarratives capable of making a proportional impact. I think the competition narrative is the source of the problem, but I understand it's difficult not to be competitive if we talk about finite resources. It's a paradigm you can't free yourself from easily. When we started working on documenta, we thought we could take everything that comes our way and build together. But eventually we had to come to terms with the fact that our time and energy are limited. And then we just have to be honest that we cannot take everyone we want, even though it would be very good for us to see other people who are building things that we can learn from—not necessarily our way, but we can relate to and have a meeting point with so it's not only our perspective. It comes back to what you said about echo chambers: you can't really know what you are looking for if you want to discover something you don't know yet. I love reading popular science for this reason: it's a source of pleasure for me. It's an awesome feeling to suddenly understand something: it really expands the horizon, nearly a spiritual experience. I see scientific work as a form of art. There is something exciting about looking for the answer in the unknown. We talked about this before—you source a lot from scientific research, don't you?

LC: Yes, it's indeed a great feeling to expand understanding for us and our work. It's also important to think about how communication works and what nonverbal tools are out there. Recently we learned about eye movement desensitization and reprocessing (EMDR), which is a therapy developed by Francine Shapiro, who discovered that moving your eyes laterally helps to resolve symptoms of traumatic life experiences. It took her quite a while to get proper attention to her discovery, but if you think about it, it totally makes sense—if you'd run away from danger, in a hypothetical prehistoric setting, you would quite likely move in actual space. As we move, walk, run, or bike, our eyes engage in lots of lateral movement, so this action is paired with a mechanism that should protect us from getting traumatized. So now, we are trying to edit a video wall in such a way that it will guide the viewer's gaze from left to right and back, and by doing so, facilitate a reprocessing of traumatic or disturbing memory.

FR: A tennis match would be good for that!

LC: Absolutely! Maybe that's why people like it so much.

FR: Exactly.

LC: And people suffering from PTSD should be given free tickets to Wimbledon.

FR: Doesn't have to be a good game—it's more important to just keep the ball flowing. So it doesn't need to be a professional match.

LC: Right, so the two players would not play against each other, but they would play together because the winner of such a competition would have to be a team, the team that kept the ball in the longest exchange. Maybe this is how you address the problem in the bones?

FR: I think what we're doing right now is like that, in small ways: to make things less competitive.

1.

In the book *Red, White & Black: Cinema and the Structure of U.S. Antagonisms* (2010), Frank B. Wilderson III asks, "Can film tell the story of a sentient being whose story can be neither recognized nor incorporated into Human civil society?" Can art?

2.

What relation does sentience have to subjectivity? What relation does art have to civil society?

Another way to put it is to track the history of what Fred Moten, riffing on Karl Marx, calls "commodities who speak." In his 1899 short story collection *The Conjure Woman*, the important African American author Charles Chesnutt conjures the past over and over again. He does this by amending the nostalgic and paternal constructs of the dialectal plantation tradition. Deploying the political economy of plantation and industry, Chesnutt's stories help to inaugurate the hauntological tradition in the black literary imagination: those non-subjects that have produced the American aesthetic imagination. I say "hauntological" to refer to some gamut of study engaging with ghosts: Avery Gordon's *Ghostly Matters* (1997), Jacques Derrida's *Specters of Marx* (1993), Toni Morrison's *Beloved* (1987) and *Playing in the Dark* (1992), Saidiya Hartman's "Venus in Two Acts" (2008). I say "hauntological" to tune into something that came before me—not quite my ancestors but rather something a little more seamless and a little less profane.

Chesnutt's story "The Goophered Grapevine," first published in *Atlantic Monthly* in 1887, begins with a white couple encountering a "shrewd" black man—Julius—who tells them a story. Julius's tale worries the married couple, who moves from Ohio to North Carolina to pursue the business of grape cultivation. For this man and his wife, the Southern states are a site not only of economic opportunity but also of emotional uplift and vitality. They boast a warm climate as well as cheap land and labor—just what the doctor ordered. This post–Civil War story is familiar today in a discourse shorthanded as "gentrification." As the white narrator named John describes the grape agriculture industry signaled in the story's title, he says, "Like most Southern industries, it had felt the blight of war and had fallen into desuetude." This destruction, decay, and abandonment opens up space for exploitative capital investment and innovation.

Upon the couple's arrival to their new home and business venture, the ghosts of the plantation stir. They stumble upon a "venerable-looking colored man," the ex-slave Julius McAdoo who was born and raised on the plantation. Julius is no ghost, but he's old and he's black and he's a trickster, so just like that, he is always and already set outside the bounds between dead and alive. Like an apparition, the otherness of Julius is marked even as the northern visitor is the unmarked, purportedly universal white man of late nineteenth-century United States. In this opening scene of encounter, Julius doubly yields the strange against the curious couple: One, Julius acknowledges that they do not know each other and thus might have reason to be suspect. When he says, "Well, suh, you is a stranger ter me, en I is a stranger ter you," he marks the gulf between himself and the "suh." Two, Julius makes it known that the vineyard is "goophered,—conju'd, bewitch'," which prompts the telling of the cautionary tale.

Julius narrates a story—a story within Chesnutt's story—set in the antebellum period. The first-person narrator (John) calls it "his memory—or imagination." Indeed! Julius warns John and his wife Annie not to buy the plantation because it is "goophered." In Julius's story of slavery, the owner of the plantation, Mars Dugal', enlists Aun' Peggy, the conjure woman, to goopher the grapevines so that the slaves will not steal and eat the scuppernong grapes. An old slave, Henry, eats the grapes because he was not informed of the goophering, and his bodily vitality begins to mimic the grape season. Mars Dugal' sets into motion a plan that moves Henry around in the market by the sale and purchase of this human commodity, buying and selling slave labor as fixed yet short-lived capital in order to increase profits.

Aun' Peggy is also doubly foreign. As Julius tells John, "She wuz a witch 'sides bein' a conjuh 'oman." She is not just a conjurer who engages in social healing practices; she *is* a witch, which is, as we learn from the likes of Maryse Condé and Silvia Federici, a label for the condemned, the accused, and the feared foreigner. Even if we are not sure of Peggy's own legal status, it is significant that Aun' Peggy lives in a cabin "down 'amongst' de free niggers." Her proximity to "free niggers" signals both a danger and a form of knowledge. Her paraplantational conjurework, adjacent to the mass labor on the profit-generating site of the American plantation, mines the border between public and private. But it also points to the co-constitution of the plot and the plantation, the slave's quarters and the Big House, or, in the particular case of Aunt Peggy, living as a slave and living next to a slave.

By troubling lines of property and propriety, Chesnutt's "The Goophered Grapevine" gives us crucial insight into the mystifying stories told in the name of profit. The economy of conjuring and its failures threaten the productivity of the plantation. The animism of Chesnutt's story is sophisticated. Trees are enlivened. Branches stretch around, emoting. Magic does not quite cut it; we might instead say his is an art of communing with ghosts.

What relation does a ghost have to the subject? The figure of the ghost operates in part to explain how a subject is formed through social power, especially at the vexed border line of enslaved and ex-slave as an already unstable category of the subject. We turn to art because it's supposed to do something different than literature, different than Hollywood, different than industry. Yet art still programs subjects into myths of freedom and unfreedom.

3.

animist chant—
a chant that wakens the soul of things
animate and inanimate
a spell that rouses the spirit
That lives in bone and stone
—Harry Garuba, *Animist Chants and Memorials* (2017)

Slavery haunts the art world.

4.

In 2014, I wrote about the arts collective HowDo YouSayYaminAfrican?, also known as Yams, for a Caribbean arts journal called *ARC*. Yams had withdrawn their film *Good Stock on the Dimension Floor: An Opera* from the Whitney Biennial in general protest of the whiteness of the exhibition and, in particular, white artist Joe Scanlan's use of the black woman character Donelle Wolford. Around six years later, while working on this essay, I was twice asked to write something about the Whitney's "Collective Actions" exhibition, an exhibition that mined prints produced for COVID-19 and Black Lives Matter fundraisers, and which was canceled not long after it was announced because of criticisms made online. Twice, I said no. Following the 2019 ousting of Warren Kanders, former vice chair of the board of the Whitney Museum and CEO of Safariland, I wanted more than seeing ourselves represented in the lush background of the fire. (Somewhere in the midst of this tornado, I saw a tweet that was like, the only Whitney I recognize: and underneath was Whitney Houston. Houston, talking about modeling in a 1995 documentary called *Scratch the Surface,* said, "After smiling for four or five hours, that whole gig, you go, okay enough. It's enough.")

Is this recent history of the Whitney yet another tale of how black people approach art as a history of nonhumans and how museums and the curatorial field often attempt to restore or that humanity?

Okay, enough. It's enough.

5.

When I started writing this essay, I had "Non-Subjects in the Art World" in mind as a possible title, or at least a theme. There it was hanging over my document: non-subject. It's not that the theme is no longer present, but negation seemed to put too much pressure on the thing that it is negating. Non-subjects refer back to subjects, lassoing them back to their reference. The same goes for the term "non-white." Yet this very kind of negative makes "anti-black racism" productive as a category.

The third volume of the journal *Propter Nos*, published by the publishing collective True Leap Press, was themed "ANTI-/NON-" and explored what the aesthetic category of the negative does for black radicalism. They asked:

> Are "anti-" and "non-" collapsible terms? Are the "anti-" and/or "non-" only thinkable in opposition to an affirmative posture; or, is there a mode of accounting for the space between, and outside of, these terms? What do "anti-" and/or "non-" offer as approaches to aesthetic practice, political action, or theoretical inquiry, rather than mere descriptors? What is the generative potential of the (perceived) passivity or resignation of orientations toward the "anti-" and/or "non-"? And need there be the promise of generativity in order to ground a politics?

We could add "de-" here. Wilderson, again, who I just can't stop reading, writes of "the Black position, who is less a site of subjectification and more a site of desubjectification."

6.

The temporality of slavery is long, showing up over and over in the present. If we think about cybernetics as having to do with the relationship between machines and the living, artists working at the intersection of technology and blackness (i.e. histories of force, histories of enslavement) really do upend those terms as given by enlarging and enlivening that history. I'm especially thinking here of American Artist, whose work explores anti-blackness in online and digital spaces. (His 2020 work *Looted* replaced all the works displayed on Whitney Museum's website with an image of plywood, reorienting what some see as art as stolen goods, as loot, while also responding to how the Whitney actually boarded up its own windows in the wake of the summer's protests.) But I might as well think here of those many, many artists in the U.S. South like Lonnie Holley, Dinah Young, Thornton Dial, and Joe Minter. Their technology is boundless.

Social media requires a kind of violent repetition: it profits from circulation after circulation of yet another murder. In response, social media distributes crumbs of agency—a post here, a like there—and calls it democratic participation. Some banner flashing up on your phone.

7.

At the end of August, the Upper West Side Open Hearts Initiative held a "sleepout and art protest" outside of the Lucerne Hotel. During the height of the pandemic in New York City, hundreds of homeless men and women were relocated to hotels on the Upper West Side in order to thin out some of the shelters, which were some of the most at-risk spaces for the spread of COVID-19. Many on the Upper West Side wanted them out. The Open Hearts Initiative's effort, they said, was "to show solidarity" with those homeless New Yorkers.

I want solidarity to be possible without the conjoining of symbolic gestures (art) and protest, not to keep it pure but to keep it paradoxical. In the United States, at the time of writing, with over five million people infected and 260,000 dead from COVID-19, a huge number of working-class people still tend to the wants and needs of people under lockdown; those affected have disproportionately been black and brown people. (With every edit to this draft, those numbers rise.)

This patronizing—and painful—metaphoric solidarity from rich white people, the kind the art world loves, made me think of the writer and artist Eileen Myles. A couple months before Donald Trump took office, Myles spent a week sleeping on the streets as part of a "street retreat." "We don't say we're homeless," they told the blog *Hyperallergic*; "we're simply living on the streets for several days, relying on the generosity of the streets to take care of us."

As Frantz Fanon wrote in *The Wretched of the Earth* (1961): "The cause is effect: you are rich because you are white, you are white because you are rich."

In these two examples of empty speech acts—the sleepout/art protest and the street retreat—the subject in subjectivity has been displaced. The white subject defines itself here against the homeless. The white subject needs the homeless for its own self-definition.

8.

There has also been, in response to people setting literal fire to their streets, the plea that their jobs could be a little less racist, a little more pleasant.

We drown, in the free wine, in the demands for subjects, for humans. There is no ethical museum, itself rotting from the outside in. This is what you came for: to fucking rage, to outdo, to rock out, to get low, to let go, to get loose, to lose it. Rage labor.

9.

People worship technology at the same time that they rebuke it. Too much screen time, we tweet. We talk about technology like god, like an unnatural power, like addicts. In an intimate encounter between physical body and nonphysical spirit, human body and spectrality, volition and coercion, and subject and subjection, we can turn briefly to the history of the human articulated by the figure of the ghost.

I raise more questions than answers. What is the subject in the wake of the plantation? What happens if we do not aim to domesticate the ghost, the non-subjects, the ex-slave? What happens if we do not drive it away or, in Toni Morrison's words, "beat off [. . .] the ghost"? The generativity of ghostliness is its otherness yet its insistence as subject position. The ghost holds some uncanny residue of the human, a lifelike specter. It is in this sense that the ghost brings into view the mutually constitutive character of becoming subject and being subjected to, being human and being barred from the humanity that needs you to exist, being a citizen and being a slave. The ghost disrupts the conclusive predication of the human by asking to be addressed. It asks us to come face to face with the ghost, to confront the ghost, to say "what up" to the ghost. If history, official public history, is a history of human subjects, the ghost articulates the unverifiable. If the history of the human is a history of sanctity and impermeability, the ghost stages a disjointing effect that leaves us seeking to make sense of the relation between self and other.

"We willfully engage in acts of self-surveillance," writes communications scholar Sun-ha Hong in his 2016 article "Data's Intimacy," as the result of a certain "technological fantasy [. . .] that machines will know us better than we know ourselves."[1] Hong was referring to seemingly innocuous devices such as FitBits,[2] but a small cohort of artists have taken this desire for intimacy via the machinic a step further: I am referring to white women flirting with the police, often via surveillance apparatuses, in the name of art. Intimacy, of course, is founded on a sort of mutual trust—in this case between surveyor and surveyed—which some groups (e.g. white women, perceived as in need of protection) experience more than others (e.g. Black men, perceived as threats). Three artists—Jill Magid, Sophie Calle, and Julia Scher—have used their position as white women to blur the line between surveillance and voyeurism in works spanning from 1981 to 2007. Though the works were meant to comment on topics like permission, desire, gender, and technology, these works are equally instructive when it comes to describing relationships between the police and white women.

In a 2011 talk at the Harvard Graduate School of Design, Jill Magid said that her early works involving romantic relationships with officers were not meant to make a statement about the police or surveillance but rather were investigations into complex and often opaque systems.[3] She made her 2002 *System Azure Security Ornamentation*, for example, after mentioning the presence of surveillance cameras throughout Amsterdam to some colleagues who in turn doubted their existence. Bejeweling the security cameras first at her studio, and later at the Amsterdam police headquarters with rhinestones with the permission of the police, she sought to draw attention to their presence. The work speaks to a pre-Snowden public consciousness of surveillance.

In *System Azure Security Ornamentation*, she employs her strategy of simply pointing out the truth to great effect. Two other works, though—*Evidence Locker* (2004) and *Lincoln Ocean Victor Eddy* (2007)—are more concerning in their neutrality, and read more as cooperation with the police than as exposure of their practices. *Evidence Locker* was commissioned for the 2004 Liverpool Biennial; at the time, Liverpool had the largest surveillance system of any English city. In her 2011 Harvard lecture, Magid speculates that the system was installed in response the city's historic riots, such as the 1981 Toxtech riots, which were a response to police brutality against Black people in the city. Magid became intrigued by the fact that anyone could request a copy of the system's footage, and so she did, framing her request forms as love letters she later compiled into a novella. She wore a red coat, told the police her location, and as she walked about the city, the police would train their cameras onto her. She then used the resulting footage to create a series of videos. In *Trust* (2004), Magid

1 Sun-ha Hong, "Data's Intimacy: Machinic Sensibility and the Quantified Self," *communication +1* 5, no. 1 (2016): Article 3, https://scholarworks.umass.edu/cpo/vol5/iss1/3.

2 Hong shows that insurance companies give them out to customers and is suspicious of their motives: though it's currently illegal in the US for companies to use that data to determine a client's "healthiness" and thus their insurance premium, Hong argues that's the horizon of potential use.

3 Jill Magid, "Embedded," artist talk, Harvard Graduate School of Design, October 24, 2011, Cambridge, MA, YouTube video, 1:20:37, at 23:20, https://www.youtube.com/watch?v=MQE2zW8yUWI&t=3849s.

4 Jill Magid, "Permission Is a Material," interview by Jeff Edwards, *ARTPULSE* 4, no. 15, 2013, artpulsemagazine.com/permission-is-a-material-an-interview-with-jill-magid.

wears an earpiece, allowing the officers to speak to her as she perambulates. At one point, she closes her eyes, and in so doing, invites them to guide her through a bustling crowd, trusting in their protection and rendering herself helpless. Magid called the process "one of the most romantic moments of my life."[4] But a protector-protected relationship always necessitates the existence of a threat. This made anyone who threatened to get in Magid's way a potential source of danger, the enemy, the other, even though they were not in on the game. She concluded the series with *Final Tour* (2004), a time lapse in which Magid rides off through the city at sunset on the back of a police officer's motorcycle.

In 2007, Magid—who is American—moved back to New York after several years in Europe to find that the city had changed rather drastically after the attacks on September 11. She began *Lincoln Ocean Victor Eddy* after hearing a certain warning announced regularly on the subway: "You may be searched." Magid decided that, in fact, she *wanted* to be searched and began *asking* officers to search her. She simply could not wait for her turn to come. The resulting work suggests that she found the prospect romantically exciting. When she finally asked a male cop to search her, he told her, "I can't search you, you're a woman," affirming that white women are seldom perceived as threats, or that the officer too saw any such interactions romantically or erotically charged and thus inappropriate. Magid remained intent on learning what these officers do on the subway all night while patrolling and managed to convince the officer who rejected her to let her shadow him during his nightly rounds. She left postcards of her favorite artworks on his car and photographed some of their interactions, never showing his face, as their on-duty relationship was strictly forbidden. In *Waiting* (2007), he's seen from across the subway platform, face turned away. The photograph is grainy due to the subway platform's lack of light. She followed him at a distance, and they often spoke in code (turned into an acronym, the work's title spells "love"). The photograph *Holding His Gun* (2007) shows her hand on his weapon, and in *His Shirt* (2007), she's wearing his navy-blue button-up as a dress. Installed in her 2007 solo show at Gagosian, Magid included a bullet from his gun in a vitrine as a symbol of their mutual trust: the bullet, which bears marks unique to his gun, could easily be traced back to him.

Lincoln Ocean Victor Eddy is probably supposed to be disturbing, and in that, it's quite successful. But I'm left feeling more disturbed about Magid's cooperation with and romanticizing of the police than I am about any revelations regarding "random" subway searches. Whereas Magid's work is concerned with permission and involves forming intimate relationships with impersonal institutions, Calle's oeuvre is firmly devoted to desire. Calle asked her mother to hire a private detective to follow her around Paris on a day of her mother's choosing, a day Calle wouldn't expect. The resulting *La Filature (The Shadow)* (1981), comprises photographs the detective took, as well as notes about his subject's movements, contrasted with photographs and notes of Calle's own. If the artist's other works are any indication, she was pleased to intrigue: the year prior, she had followed an attractive man named Henri B. to Venice, taking similar photographs and notes documenting his whereabouts. To be followed and watched, for Calle, is to be an object of desire. As John Berger put it in his famous *Ways of Seeing* (1972), women learn to "watch themselves being looked at" as a result of a patriarchal society obsessed with looking at women. Viewing *La Filature*, it is clear that Calle isn't being investigated as a threat or as a suspect, and not only because she initiated the whole ordeal. Calle knows that normative scripts of heterosexual desire will make us read this work as a love story. She plays into the misogynistic idea that women should feel flattered when followed.

In the most convincing feminist reading of Calle's practice, Anna Watkins-Fisher calls the artist's approach a form of "parasite feminism," comparing it to the work of autofiction pioneer Chris Kraus.[5] Parasite feminism, Watkins-Fisher writes, involves knowingly exaggerating the misogynistic stereotype of women as needy and dependent to the point that she becomes no longer passive but threatening. In other words, the feminist artist feeds on the patriarchal host—needs him to survive—needs him so much that he's sucked dry, has nothing left, then dies. Here, Calle preys upon law enforcement's perception of white women as objects of innocence in need of their protection, but in engaging in this dynamic willfully and knowingly, she endeavors to position her perceived passivity as a type of power.[6]

As Watkins-Fisher notes, works that take parasite feminism as a strategy tend to seem, at first, to center the patriarchal host, but are in fact about the parasite herself. *La Filature*, *Evidence Locker*, and *Lincoln Ocean Victor Eddy* all reflect more about the position white women enjoy vis-à-vis the police than offer insight into policing at large. Calle, like Magid, seems to perceive her whiteness as neutral rather than the condition that makes such work possible.

Should you need convincing that the police prioritize the protection of white women, especially when it enables them to justify their perception of Black men as threats, allow me to remind you of two among many scenarios that make this plain. Emmett Till, you may recall—a fourteen-year-old Black boy from Missisippi—was lynched in 1955 after a white woman accused him of flirting with her. And in the state of Minnesota, the only officer to have ever been convicted and sentenced for killing a civilian was Mohamed Noor, a Black officer who shot a white woman named Justine Damond in 2017. The white Minnesota officer who shot Philando Castile on video in the year prior, meanwhile, was acquitted. I hope, however, that this

5 Anna Watkins-Fisher, "Manic Impositions: The Parasitical Art of Chris Kraus and Sophie Calle," in "Viral," ed. Patricia Clough and Jasbir Puar, special issue, *WSQ: Women's Studies Quarterly* 40, no. 1/2 (Spring/Summer 2012): 223–35.

6 This is only so powerful, as when the host dies, the parasite dies with it—unless she is able to find another host to feed her.

7 Julia Scher, artist talk, Art Institute of Chicago, March 29, 2017, YouTube, 42:08, at 4:07, www.youtube.com/watch?v=k-Gs_9KszOQ.

8 Magid and Scher were both at MIT's Visual Arts Program, which is now the Art, Culture, and Technology department, in the late 1990s and early 2000s.

9 I reviewed Scher's 2019 restaging of *Wonderland* at Esther Schipper, Berlin, for *Art in America* and am recycling some ideas from that review here. See Emily Watlington, "Julia Scher," *Art in America*, April 1, 2019, https://www.artnews.com/art-in-america/aia-reviews/julia-scher-62663.

10 Julia Scher, "Taping Thousands of Hours of Parking Lots, Just to Be Sure," interview by Maurin Dietrich, *Mousse Magazine* 62, February/March 2018.

information is out of date by the time this book is published, as we eagerly await the results of trials involving George Floyd and Daunte Wright.

Against this backdrop, it is concerning that Calle romanticizes her privileged protection, even as it's unclear—as it is throughout her oeuvre—how critical or complacent she is as she reveals patriarchal and heteronormative scripts that underpin our understanding of desire. Magid, similarly, avoids taking firm stances in her early works. While Julia Scher too avoids pointed criticality, her approach is more absurdist than it is romantic. Scher has said she is less interested in, say, exposing the horror of the CIA than she is in lip-synching.[7] A contemporary of Calle and a professor of Magid's, her technology-centered work often upends the binary of protector and protected along the lines of age and gender.[8] Like *La Filature*, much of Scher's work is concerned with private security companies rather than law enforcement. The installation *Wonderland* (1997), for instance, includes a miniaturized office space for watching surveillance footage, scaled to put children in the position of power.[9] At Esther Schipper in Berlin in 2019, *Wonderland* was expanded to include large-scale photographic portraits of children (including a young Lena Dunham) in pink Security by Julia uniforms. Candy was splayed along the countertop, and it occurred to me that a child wouldn't think twice about grabbing some, whereas I felt unsure if the candy was part of the art and unsure if I was being watched. This experience testifies to Scher's sharp commentary on learned behavior. *Discipline Masters* (1988), meanwhile, is a confessional video in which Scher exploits the camera's capacity for furnishing evidence. She talks about her life since childhood, but, typically screened in a gallery, I imagine few viewers endure the entire four-hour duration. The artist bares all, bares too much, to the point that it's boring, discussing, for instance, how her mother forbade her from eating slices of bread from the middle of the loaf. The artist's *Mousse* interview with Maurin Dietrich is titled "Taping Thousands of Hours of Parking Lots, Just to Be Sure"—a nod to the idea that we collect more data than we can possibly analyze.[10] Most of it is extremely boring. But perhaps—the logic seems to go—if we collect anything and everything, we can find the needle in the haystack. Scher gives her life story away before it can be taken from her and fights surveillance with boredom. Her strategy is quite the opposite of turning herself into an object of intrigue or desire, and her non-romanticized approach is a welcome relief. But still, I'm afraid, her strategy is best suited for white women alone. Seemingly minor, boring details can still result in death sentences for Black men—an illegally hung air freshener was cited as a reason for the traffic stop that resulted in the murder of Daunte Wright.

These works confront myriad systems of power—private security, the police, patriarchy, cybersecurity, heteronormativity—but there's one they all neglect to name: whiteness. I myself was certainly not interrogating the privilege I experience as a white woman encountering law enforcement the years these works were made. That came only later, thanks to the activism of Black Lives Matter and other Black organizers. Still, it is not anachronistic to consider these works through an abolitionist or reformist lens, as calls for abolition or reform, as well as critiques of police brutality, are far from new. Because the police are not actually the primary objects of critique in any of these works, they reveal unconscious cultural attitudes about the roles of policing vis-à-vis white women.

Taken together, these works, to varying degrees, suggest that white women might tool their position as innocent objects of intrigue into a sort of power. But it matters greatly what we use that power for. The work of all three artists reads as deeply aware and critical of the gendered relationships in which they are engaging (no doubt informed by lived experience), as well as the gendered rules they are, to varying degrees, playing by, albeit with a knowing wink. Their whiteness, however, is approached as though it were neutral or universal. My concern isn't simply to say these works are not abolitionist or reformist—they aren't, but again, a lot of things weren't then, and aren't now. It's rather that these works reveal a sense of powerlessness, or a learned helplessness—the feeling that there is no choice but to submit, so we might as well try and adapt existing rules to our advantage. I understand this kind of learned helplessness when faced with patriarchal norms as a feminist strategy in many scenarios. The police, however patriarchal they may be, are still a concrete institution that can in fact be abolished. Besides, adapting the rules of policing to the advantage of white women can only happen at the expense of Black men.

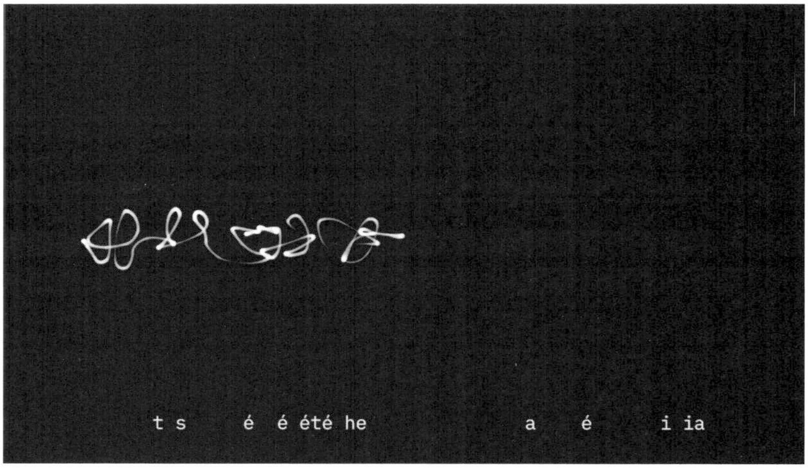

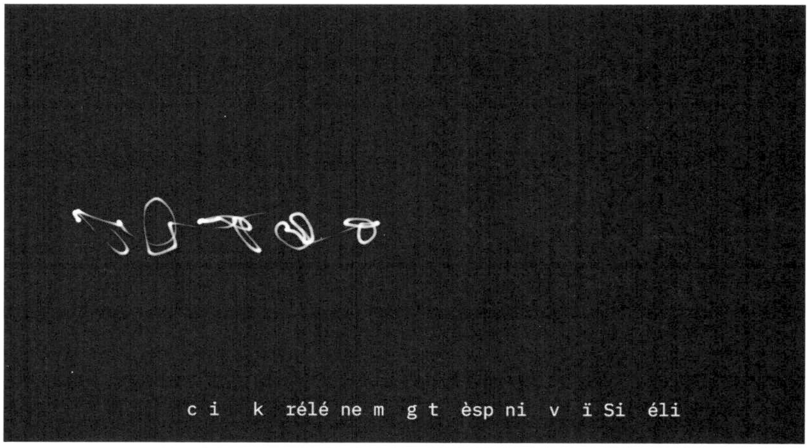

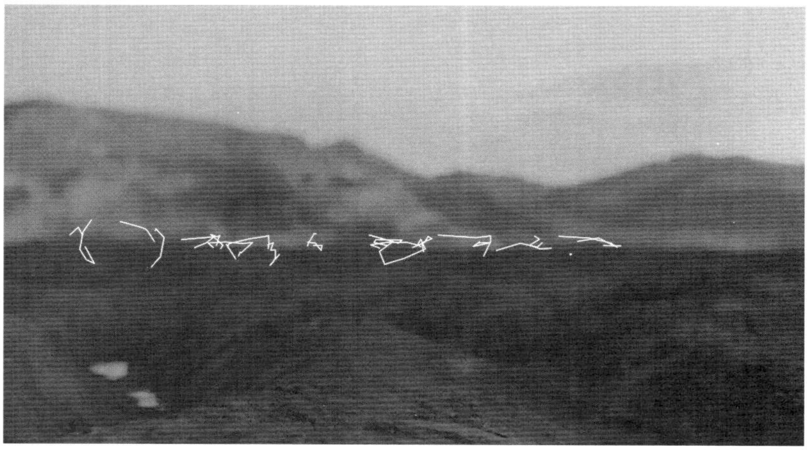

Jenna Sutela, *nimiia cétiï*, 2018. Video stills

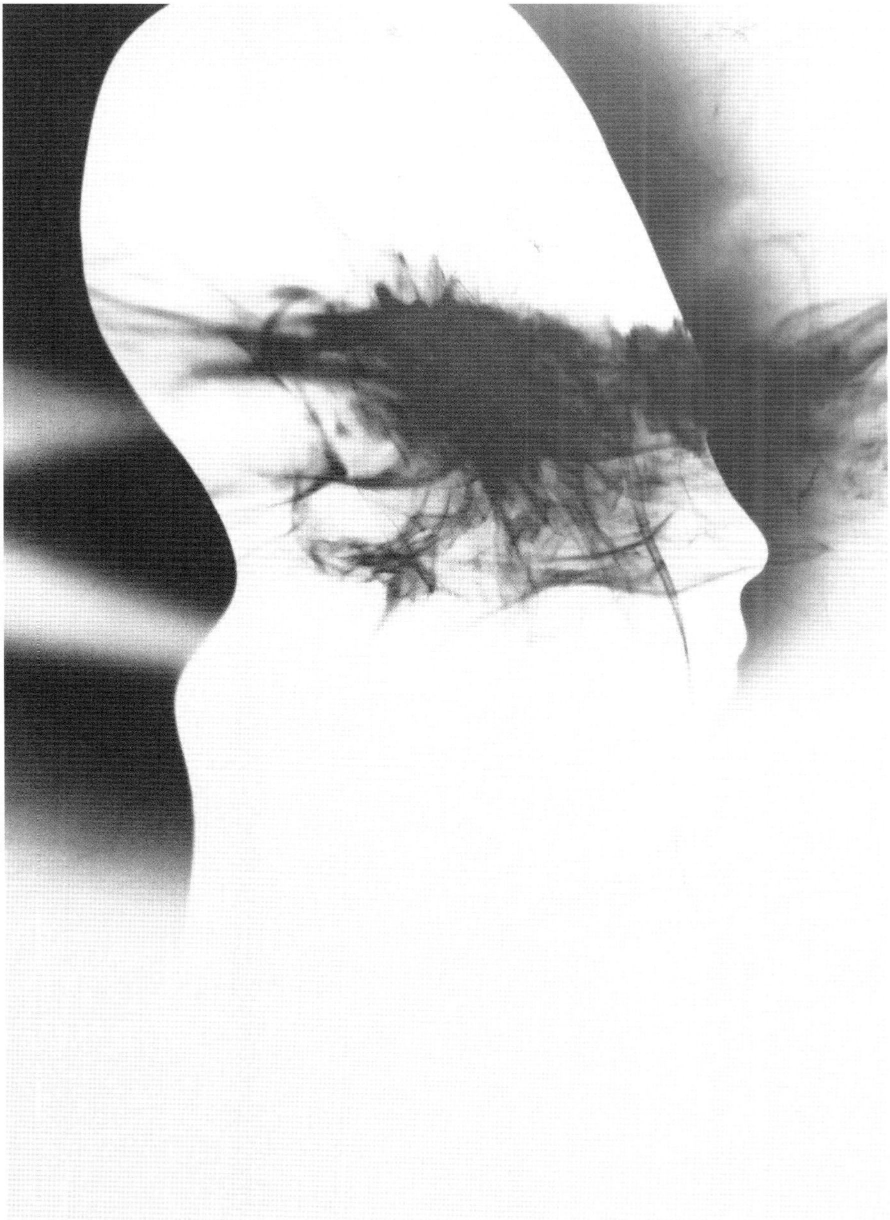

Jenna Sutela, *The figure of the deep*, 2020. Photogram

Jenna Sutela, *I Magma*, 2019, and *Early reign the lunar goddess*, 2020, in the exhibition "NO NO NSE NSE"

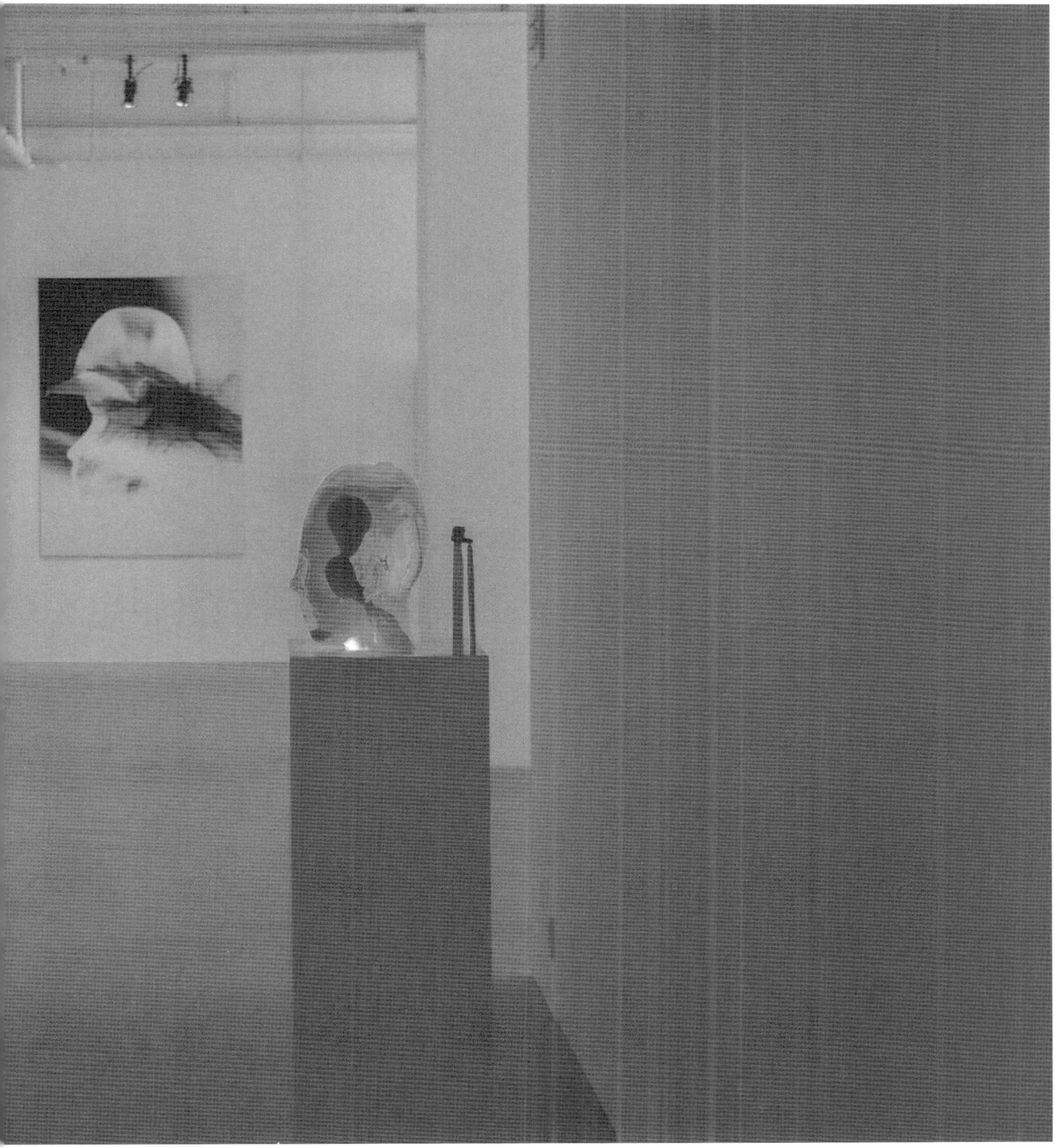

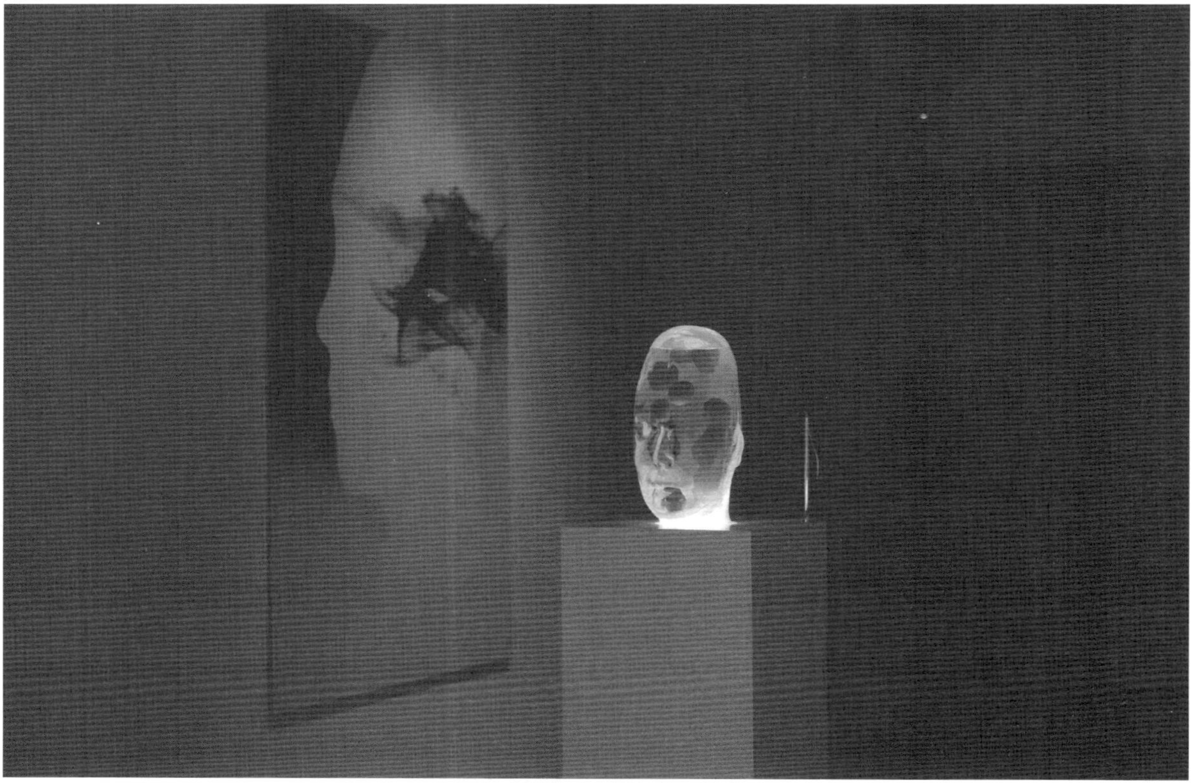

Jenna Sutela, *I Magma*, 2019, and *They saw the ear world*, 2020, in the exhibition "NO NO NSE NSE"

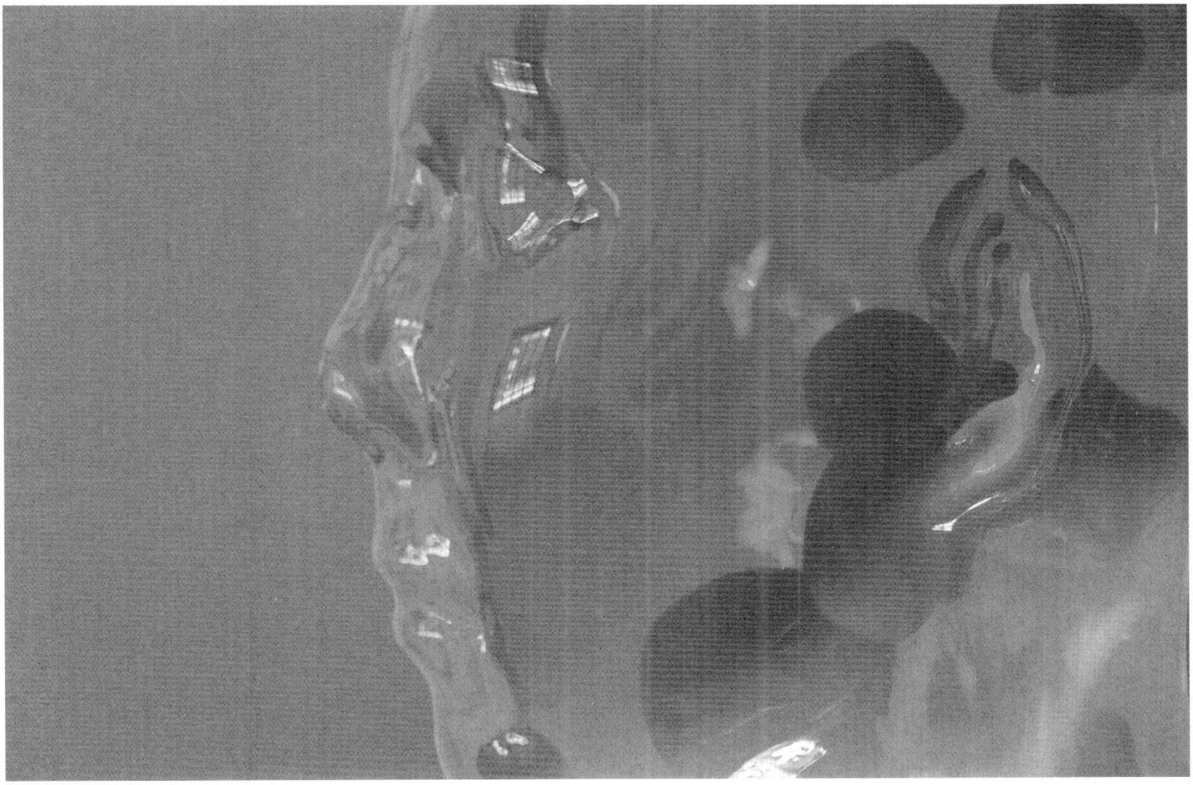

Jenna Sutela, *I Magma*, 2019

There is no interpretation that does not involve a possible mistake.[1]
—Georges Canguilhem

In his mid-2000s video work, Takeshi Murata digitally reprocesses film to create an array of uncanny effects. In *Monster Movie* (2005), he loops short excerpts from a 1981 comedy flop called *Caveman*. The head of a hairy beast rises from the surface of water. The monster dances and lumbers around a cave. But only the movements repeat. With each loop, a new set of colors takes on the monster's choreography. It's a familiar aesthetic, a combination of pixel-like chunks that blend and curve in a hallucinatory visual mess. It's an elaboration of the ever-so-brief digital sludge that appears on streaming video during a lag in the latency of internet connectivity.

This piece has become a landmark of the glitch technique called "datamoshing." Murata unlinks two kinds of data involved in video compression: one about movement and one about content.[2] This produces a kind of digital impressionism that dithers on the edges of legibility. In *Untitled (Silver)* (2006), the information mismatch is slowed, melting the black-and-white images of Barbara Steele in Mario Brava's 1960 gothic horror *Black Sunday*. In *Untitled (Pink Dot)* (2006), the machismo of Sylvester Stallone in *Rambo* plays a distorted hide-and-seek with a bright pink sphere. As movement form is painted with a palette of out-of-sync content, we wonder what is missing, and whether the effect is a kind of digital wear or lossiness. Datamoshing cleverly collapses the etiologies of distinct glitch forms.

Watching these works in late 2020 is cathartic for reasons Murata could not have imagined. The intentional glitching of form and representation invites us to linger in a space we tend to look past in frustration. Though these works confound any real attempt to distinguish possible latency glitches from display errors from other resolution problems that might not be built into the work itself, we are relieved of the duty to localize the fault. We are released from the tensed-up worry about stutters and lags. Against this ambient affect in a world of remote access, Murata's datamoshing features as painterly and soothing and surprising.

Our embodied frustration of the glitch encounter is an invitation to awareness at the points of contact between our bodies and our machines. Beholding a beautiful glitch turns the stuttering image of a colleague into a dance, the tonal slog of audio delay into a song. These are unwitting collaborations at the human-technology hinge.

When we talk about the glitch, we usually hold stable the body being mediated. It's the technology that acts up. But what of the vocal stutter assumed to be caused by a glitch in a phone call or video conference software? What about the difficulty in locating the noise of a headphone microphone tapping against a tremoring body? If we're training our focus

1 Georges Canguilhem, *On the Normal and the Pathological*, trans. Carolyn R. Fawcett in collaboration with Robert S. Cohen (Dordrecht, Holland: D. Reidel Publishing Company, 1978), 277.

2 These two kinds of data compress image information by transmitting only aspects of an image that have changed from frame to frame, reducing an otherwise heavy transmission load of each frame received anew. Lags in connectivity can often have the unintended effect of briefly separating movement data from the images, producing a datamoshed flash.

3 Andrew Brooks, "Glitch/Failure: Constructing a Queer Politics of Listening," *Leonardo Music Journal* 25 (2015): 37.

4 Jonathan Sterne, "Compression: A Loose History," in *Signal Traffic: Critical Studies of Media Infrastructures*, ed. Lisa Parks and Nicole Starosielski (Chicago: University of Illinois Press, 2015), 31–52.

5 Hito Steyerl, "In Defense of the Poor Image," *e-flux Journal* no. 10 (November 2009), https://www.e-flux.com/journal/10/61362/in-defense-of-the-poor-image.

6 Peter Krapp, *Noise Channels: Glitch and Error in Digital Culture* (Minneapolis: University of Minnesota Press, 2011).

7 Mara Mills, "Deaf Jam: From Inscription to Reproduction to Information," *Social Text* 28, no. 1 (Spring 2010): 35–58.

on the glitch as an illuminating moment, could we understand it the way we understand errors of the body? Can we speak of disabled machines? Or should we say "machines with disabilities"?

Glitch as Redemption

What we could call glitch theory has moved on the possibility of using the glitch as a conduit for delimiting the protocols of systems whose otherwise finessed mechanisms conceal themselves in normal function. The general tendency of these works to redeem the disruption of the glitch is captured well by Andrew Brooks's 2015 queering of the glitch as a sonic aesthetic: "Claiming aberration and failure as sites of creation, glitch artists and musicians create unstable systems that allow for emergent and distributed forms of agency."[3] Elaborated as a disruption in anything we might call a system, the glitch is now a discursive object that stands in for a host of meanings about failure, distortion, wear, noise, error, and more.

This tendency performs a powerful check on innovation-centric, pioneer-oriented discourses about technology. As Jonathan Sterne has pointed out, media scholars often lionize a notion of innovation as the elimination of "noise" and insist on unreliable metrics of verisimilitude and signal definition. The supposed match between a being-there experience and the hearing-it-recorded experience, in the case of audio compression, takes precedence over what Sterne notes is a host of intensities and intimacies that are predicated on the *lack* of verisimilitude or definition.[4] The crackles and hisses on a dusty or scratched vinyl record, the bootleg aesthetics of a surreptitiously copied VHS tape, the unavoidable scrambling of a radio program on a road trip are all occasions when a medium's crack is where its light gets in.

Hito Steyerl's 2009 defense of the "poor image" focused on productive infringements of the possibility of high resolution, remains important. Fugitives of the commercial behemoth of intellectual property protection have a raggedy endurance that shows more in what it can't show. The poor image names the thrill in refusals of capitalism's braid of fidelity and innovation.[5]

Glitch theorist Peter Krapp has observed that "noise channels" were crucial to the development of information theory, and notes the utility in attending to the glitch response as a conduit for the significance of communication technologies.[6] Mara Mills calls disability the "assistive pretext" in engineers' journey toward innovation, naming both the quick retreat from disabled people's actual access needs and the way disability serves as the proving ground for technical projects.[7]

The elaboration of the glitch reaches its peak in Legacy Russell's *Glitch Feminism* (2020), which seeks to reapply what she calls "*glitch-as-error*" from digital environments to the "AFK [Away from

Keyboard] world." "The glitch is a passage through which the body traverses toward liberation, a tear in the fabric of the digital," she writes.[8] "Glitch is all about traversing along edges and stepping to the limits, those we occupy and push through, on our journey to defining ourselves."[9] The glitch abstracted to such an extent loses its hold on the primordial frustration about the disruption itself, so released from the concern that it is no longer obligated to ask after the very trouble it started with. The glitch's new status as constitutive of media systems might have had a curiously docile-izing effect.

Despite the blunt strokes with which it paints similarities between digital and human domains, Russell's work signals the centrality of value operations in systems-thinking. The glitch is a designation, not an ontological status. If a computer virus is a metaphor made legible by human virality, we might well ask if we can indeed conceptualize it in terms of error at all. A virus is only a glitch if what we're trained to need to protect is a continuity in the human body; the virus does quite well for itself. Thus the glitch articulates a relationship of values between parts. As Safiya Noble notes, failures of imagination in digital design that encode racist and sexist values in search algorithms are quick to be called glitches.[10] This is not because they are spurious technical blips, but because the neutrality of a system must be nullified or be threatened.

But the word itself? *The Oxford English Dictionary* classifies the etymology as "unknown," but cites an early use in "astronaut's slang" as "a hitch or snag; a malfunction," referring, as do other dictionaries, to John Glenn's contribution to *Into Orbit*—a 1962 text written by the seven members of the U.S.'s first human spaceflight program, Project Mercury. There, Glenn wrote:

> A term we adopted to describe some of our problems was "glitch." Literally, a glitch is a spike or change in voltage in an electrical circuit which takes place when the circuit suddenly has a new load put on it. You have probably noticed a dimming of lights in your home when you turn a switch or start the dryer or the television set. Normally, these changes in voltage are protected by fuses. A glitch, however, is such a minute change in voltage that no fuse could protect against it.[11]

The astounding array of referents that get condensed in the glitch could date back to the Yiddish word *glitsch* meaning "slip" or "lapse" or "error," perhaps a derivative from the German verb *glitschen* meaning "to slip."[12] Others suggest that the term was first used to describe slip-ups on the radio—human, not machine.[13] Still others claim that the glitch was not itself a word but an acronym, for either "gremlins loose in the computer hut" or "gremlins lurking in the computer hardware." Harking to depictions of mischievous creatures who

8 Legacy Russell, *Glitch Feminism: A Manifesto* (New York: Verso, 2020), 30.

9 Russell, *Glitch Feminism*, 13.

10 Safiya Noble, *Algorithms of Oppression: How Search Engines Reinforce Racism* (New York: NYU Press, 2018), 10.

11 John Glenn et al., *Into Orbit* (London: Cassell, 1962), 245.

12 Leo Rosten, *The Joys of Yiddish* (New York: Penguin Books, 1971), s.v. "glitch"; and William Safire, "On Language," *The New York Times*, March 23, 1980, 3, quoted in Ben Zimmer, "The Hidden History of 'Glitch,'" Word Routes, Visual Thesaurus, November 3, 2013, https://www.visualthesaurus.com/cm/wordroutes/the-hidden-history-of-glitch.

13 Zimmer, "The Hidden History of 'Glitch.'"

14 Emphasis in original. Georges Canguilhem, "Machine and Organism," trans. Mark Cohen and Randall Cherry, in *Incorporations*, ed. Jonathan Crary and Sanford Kwinter (New York: Zone Books, 1992), 55.

15 Canguilhem, *On the Normal and the Pathological*.

have sabotaged the British Royal Air Force since the 1920s, the gremlin theory of the glitch renders otherwise inexplicable technical errors mythological.

As helpful as it would be to sort the cluttered discursive landscape into a careful typology of the glitch, such a typology would tell us less than the fact of the clutter itself. What is more interesting is the incitement to the glitch, the way it only sloppily captures anything necessary for repair. And if the glitch is the body of a writer, is a literary trope, is a sonic aesthetic, it becomes important for us to more precisely consider the ways that the glitch works at the interface of the human and the machine.

Errors & Organs

In 1947, French philosopher Georges Canguilhem delivered a lecture called "Machine and Organism," in which he outlined the long legacy of a "mechanistic" theory of the body whereby human life is explained in terms of machines. Often imagined as the central conceit of Cartesian innovation, Canguilhem divined an earlier tradition in Aristotelian and Kantian philosophy that might explain the theory's endurance. Canguilhem's concern was reversing the flow of the homologies between the body and machine.

It was not the body that was to be properly understood as a machine, Canguilhem offered, but machines that should be understood as extensions or organs of the body. He argued that mechanistic approaches to understanding the human form shortchange the purposiveness of all mechanisms, a purposiveness that indexes the body. "*Machines can be considered as organs of the human species*," Canguilhem writes. "A tool or machine is an organ and organs are tools or machines."[14] Canguilhem, for whom life was Life, saw the possibility of elucidating a machine's malfunction in terms of vital agency.

His 1947 lecture is of a piece with his more well-known project of historicizing modern medical and social philosophies of the body in a quantitative variation between the normal and the pathological.[15] Canguilhem insists that we could better understand the fundamentally qualitative dimensions of living, a claim he knew would smack of an atavistic call to the "humors" of the body. What he proposes, though, would remain radical for decades—and perhaps explains how, for all this, he remains an underrecognized forebear of some of the most important currents in late twentieth-century theories of media and technology.

Canguilhem offers what he calls "biological normativity" as a way of conceptualizing the relationship between an organism and its environment. Norms are not produced once and for all on the level of the social; they are located in the body's active negotiation with its surroundings. This is not reducible to a theory of adaptation, he stresses. The dynamism he specifies is in the way both the body and its ecology are mutually designed to produce a silence of the organs each individual body can understand as normal for itself.

When Canguilhem revisited *The Normal and the Pathological* in 1963, writing with the privilege of having watched the cybernetic paradigm install itself, he added several sections to the book that articulate an exquisite philosophy of error. Though he developed it by contemplating the logical inconsistencies of prevailing theories of genetic predisposition to enzymatic malfunction, he explicitly deploys the term as polysemic in order to address what he calls the "affectivity" of the discourse of disease. "To be sick is to have been made false," he writes, "to be false, not in the sense of a false bank note or a false friend, but in the sense of a 'false fold' [i.e., wrinkle: *faux pli*] or a false rhyme."[16] In his attention to the ways that medical language becomes social metaphor, he attends to what disability activists would later bring a civil rights movement to bear on.

> To be sick is to be bad, not as a bad boy but as poor land. Disease is no longer related to individual responsibility; no more imprudence, no more excess to incriminate, not even collective responsibility as in the case of epidemics. As living beings, we are the effect of the very laws of the multiplication of life, as sick men [*sic*] we are the effect of universal mixing, love and chance. All this makes us unique, as has often been written to console us for having been made from balls drawn by lot in the urn of Mendelian heredity.[17]

Error is constitutive of Life the way disability is constitutive of normalcy. Canguilhem nearly says as much:

> In the long run a malaise arises from not being sick in a world where there are sick men. And what if this were not because one is stronger than the disease or stronger than others, but simply because the occasion has not presented itself? And what if, in the end, when the occasion does arise, one were to show oneself as weak, as unprepared as, or perhaps more so than others? Thus there arises in the normal man an anxiety about having remained normal, a need for disease as a test of health, that is, as its proof, an unconscious search for disease, a provocation of it. Normal man's disease is the appearance of a fault in his biological confidence in himself.[18]

For Canguilhem, disease is something that is dangerous to cure. And when he writes that machines are organs of the body, his notion of error aptly slips between both domains—or helps muddle the distinction altogether perhaps better than Haraway's later

16 Canguilhem, *On the Normal and the Pathological*, 276.

17 Canguilhem, *On the Normal and the Pathological*, 278–9.

18 Canguilhem, *On the Normal and the Pathological*, 286.

19 This is the critique Alison Kafer offers of Haraway's cyborg and is a good summary of the reasons disability theorists are often averse to seeing the cyborg as liberatory possibility. See Alison Kafer, *Feminist, Queer, Crip* (Bloomington: Indiana University Press, 2013).

20 Michael Hagner, "Bilder der Kybernetik: Diagramm und Anthropologie, Schaltung und Nervensystem", in *Der Geist bei der Arbeit. Historische Untersuchungen zur Hirnforschung* (Göttingen: Wallstein, 2006): 195–222, quoted in "Normal and Pathological Humanity: Michael Hagner on Canguilhem," interview by Caroline A. Jones, *Thresholds* 42 (2014): 103.

21 Victoria Ann Lewis, "Crip," in *Keywords for Disability Studies*, ed. Rachel Adams, Benjamin Reiss, and David Serlin (New York: NYU Press, 2015), 46.

22 Robert McRuer, *Crip Theory: Cultural Signs of Queerness and Disability* (New York: NYU Press, 2006), 197.

23 Carrie Sandahl, "Queering the Crip or Cripping the Queer? Intersections of Queer and Crip Identities in Solo Autobiographical Performance," *GLQ* 9, no. 1–2 (2003): 51.

notion of the cyborg. Canguilhem also foregrounds what Haraway neglects: that cyborg relations are defined by practice, not the superficial political reading of a disabled body that risks reinscribing a staid medical ontology.[19] If it's true, as Michael Hagner writes, that the uptake of the cybernetic model after 1945 was enabled by a machine-centric universalism seen as an alternative to the organism-centric violence of eugenics,[20] then Canguilhem points to a model for steadying the tacks of the course.

A return to Canguilhem reels in the wild abstraction of body-machine relationships in glitch discourse. It is not that a disruptive body could be understood as a glitch in a system-like ecology, but that a glitchy machine is already a component of a vital ecology. The correction is high stakes: it compels an otherwise skin-bound politics of care for technology, it extends our obligation to cultivate Life where only inert matter was thought before, and it lodges the glitch as a permanent and necessary condition of wholeness.

Those familiar with the political conceits of disability rights and justice may recognize a kindred perspective here. In the politics and culture of communities since the 1970s in the US and the UK, disability has been conceptualized as a natural part of the human experience and not something adequately captured by individuated quests for diagnosis and cure. In disability politics, interdependence is an ideal but also factual description of the world, including collaborations between species and technologies.

The word "crip" entered disability culture first as part of the disability rights movement, then as a critique. As Victoria Ann Lewis has noted, "'crip' gained wide usage as an informal, affectionately ironic, and provocative identification among people with disabilities" in the 1970s.[21] But as disability studies grew as an interdisciplinary humanities field in the nineties, the capacity of the term to unsettle drew the outlines of a political agenda that moved beyond single-issue rights activists' focus on law and legislation and into a radical centering of disability as a living expertise that accommodated nondisabled norms, not the other way around. The word became a methodological verb for disability studies: to crip a structure, like queering it, is to show its contingency on ideologies of ability. When crip became theory in Robert McRuer's well-cited work, it came to "mark a critically disabled capacity for recognizing and withstanding the vicissitudes of compulsory able-bodiedness."[22] Crip theory made possible analysis of disability as central to the development of neoliberalism writ large. In fact, in one of the first flagship articles to delineate a specifically "crip" theoretical orientation, Carrie Sandahl refers to disabled performance artists as "glitches in theory."[23]

Crip politics is the anti-assimilationist program that challenges disability's persistent elimination from the category of human itself, by turns because disability is imagined as a life not worth living and therefore closer to an inorganic existence or because it was categorized in eugenic terms as closer to animality or objecthood. Crip is a reaction to being seen more as a technology than a user. Crip theory names, similar to its queer sibling, a universalizing tendency toward disability. And it centers disability not just conceptually but in form and practice, which means that perhaps an attempt to explain the crip orientation is itself a flight from it.

Crip is the kind of thought that asks how an anthology is like an institution and whether we should get the fuck out. Crip is the cleared schedule the day after an essay is finished, the time to recoup incorporated into the time it takes to produce. Crip is the laptop battery charge rendered as spoons. Crip is the smile of self-recognition when the glitch appears.

Inter-corporeal Glitch

A video opens with the sound of a rushing liquid. There is a steady, faint heartbeat. Pale images—they appear to be of websites, pasted on top of each other like a collage—have a general primrose color with swathes of red. A woman with a British accent starts to speak. She says, "Can blogging and social networking be positioned as a form of life, a vital practice of becoming-with, in which pulsations of distributed bodies, keyboard strokes, flickering particles of electricity, and under-ocean cables constitute a different form of tissue culture?" She continues speaking and the ambient sounds of liquid and a heartbeat continue as well. The images that make up the collage come and go, giving the impression that the whole thing is moving slowly.

The classic journalistic quip that "if it bleeds, it leads" is about mainstream news headlines and their enduring affinity with harm and destruction. In her video project *If It Reads, It Bleeds*, the artist and journalist Joanna Zylinska reverses the phrase and also takes it literally. It is not a living or bleeding body that precedes media but mediation that forms a living body. This body is of an individual, as Zylinska cites Foucault in her voice-over figuring the body of a reader and writer constituted in the process of reading and writing. But the body is also larger than the individual, one dynamic part of a "tissue culture" that connects matter, minds, and media.

This body is not only alive in this video; it is bleeding. The colors of the video shift from white fleshy tones to blood red to coagulated darkness. Zylinska produces skin, blood, and scabs as she technically sutures together images of various blog posts into a two-dimensional yet dynamic collage that seems, in the end, eerily lifelike. The twinning of biology and technology is reinforced by the heartbeat in the soundtrack that drifts in and out of our focus but never quite disappears, reminding us that when we speak of technology, we often forget the embedded pulse of biology. At one point in the video, there are words, covered over by other capillaries, that read: "Have you ever looked down at your body and been surprised by what

24 See, for example, Jane Bennett, *Vibrant Matter: A Political Ecology of Things* (Durham, NC: Duke University Press, 2010); and Karen Barad, *Meeting the University Halfway: Quantum Physics and the Entanglement of Matter and Meaning* (Durham, NC: Duke University Press, 2007).

25 Sarah Kember and Joanna Zylinska, *Life After New Media: Mediation as a Vital Process* (Cambridge, MA: MIT Press, 2012), 14.

26 Emphasis in original. Kember and Zylinska, *Life After New Media*, 17.

27 Quoted in Julia Watts Belsar, "Vital Wheels: Disability, Relationality, and the Queer Animacy of Vibrant Things," *Hypatia* 31, no. 1 (Winter 2016): 9.

is there?" The out-of-body-ness registered in this banal blog post heading is meant to remind of us of the becoming-with of our media *and* our bodies, that for however much we want to insist on the skin-bound nature of human beings apart from the cold, hard matter of our technology, we nonetheless glimpse the two together in moments when we see our bodies as objects. Or conversely, we glimpse the subjectivity of *all* matter when we forget momentarily that our earphones are not naturally existing corporeal devices.

We might read this project as an open vein into the liquid realm of what Zylinska, in her co-authored book with Sarah Kember, calls "mediation." In *Life After New Media* (2012), Kember and Zylinska inaugurate a vitalistic media studies, building on recent work in vital materialism by Jane Bennett and Karen Barad, among others.[24] Where "media" are understood as technological artifacts like a camera or its photographs, "mediation" is a more dynamic force that dips not only into the "liveness" of media (following the many studies in communication dedicated to understanding media in step with a linear temporality), but also into a "lifeness" of media.

Kember and Zylinska borrow heavily from Bernard Stiegler (who borrows heavily from Heidegger) in describing the "originary technicity" that names how humans are not in fact set apart from technology, but share a *technical tendency* that produces "our way of being-with and emerging-with technology."[25] Heidegger's claim that "the essence of technology is by no means anything technological" prefigures this "originary technicity" by signaling how technology is more "revealing" of the potentiality of matter than it is instrumental. Kember and Zylinska take up this way of considering technology as a vital process—a process reaching even to a biological sense of life—in order to show how humans become part of a complex web of interactions with nonhuman actors.

"Originary technicity," they write, "can thus be understood as a condition of openness to what is not *part* of the human."[26] Julia Watts Belsar, in her rendering of queer and crip animacies of wheelchairs with their users, suggests that this is an epistemological orientation disabled people have had long before new materialism discovered it. She finds an erotics in the writing of an anonymous blogger: "I can stroke the wheel with my fingers—it's a sexy feeling. And when I do it, I feel every last ridge of the tyre as it passes under my fingers. My fingers are the things to watch; they love and caress the wheel as it rolls underneath my hands."[27] What quivers when one side or the other of this vital connection snags or stops working?

The glitch has perhaps never structured such a collective experience, as we become our image every day for remotely accessible meetings, classes, and events. 2020 might mark a new ascendance of glitch discourse that risks regenerating the rigid distinctions between machines and organisms. Can we secure the glitch as dys/function of the organs?

John Miller

In 1988 Haim Steinbach appropriated an advertising slogan from a French telecommunications company for a work that was to be a wall text: "Tant qu'il aura des petits creux," which means, "As long as you'll have a free moment." The slogan suggests that if you have a quick break, you can squeeze in a telephone call. As Steinbach parsed it, "So long as there's a little hole, a little absence, a little vacuum". . .[1] In short, the phone company proposed to insert its services into those otherwise fleeting interstices not overtaken by work. This drive to exploit residually untapped resources laid the grounds for what would become a growing tension between technology and personal space, between what is public and what is private.

Today, phoning someone is something of an intrusion. Most prefer to text. In Bret Easton Ellis's novel *American Psycho*, published three years after Steinbach's wall piece, one finds the coldest yuppie putdown of all time: "I'll fax you an apology." Now, its sarcastic intent is unclear. The fax machine is all but forgotten. To receive such a message as a printout would seem, if anything, polite.

Curiously, the phone company's impetus to fill every available little hole extends the logic of multiplexing, which, because speech mostly consists of silent gaps, can aggregate multiple signals into a single data stream. This technique allows a single telephone line to carry several conversations. Just as multiplexing hollows out speech for economy's sake, so the telecommunications corporation hollows out leisure for the sake of profits. This inward turn to ever-smaller, leftover spaces characterizes the postcolonial era. Rather than conquering new territory, capital must exploit the interior, including especially the private life of the individual. With this, the drive toward monetization takes on a more intimate dimension. Conversely, intimacy becomes, in a way, less intimate.

Intimacy is typically defined as a close relationship between two or more individuals. As such, it is intersubjective. Nonetheless, an intimate relationship is, by definition, an *unmediated* relationship: it involves direct contact, openness, and vulnerability. Intimacy is also a euphemism for sexual relations, in short, the suspension of boundaries. Boundaries suggest that intimate space is a space that one inhabits and controls, where, outside the purview of societal expectation, one can be oneself. To some extent, this aspect of intimacy registers the existential opposition that surrealist Georges Bataille drew between the heterogenous (that which exists only for its own sake) and the homogenous (that which exists for others).[2] In other words, it hinges on the question of instrumentalization. If one is instrumentalized, if one is just a means to someone else's ends, does this preclude intimacy? If so, the contemporary mediascape tends to minimize intimacy—or at least intimacy as we once knew it.

1 Haim Steinbach, "Side Order of Facts," interview by John Miller," CAPC Bordeaux 1988, unpaginated.

2 Georges Bataille, "The Psychological Structure of Fascism," in *Visions of Excess: Selected Writings 1927–1939*, ed. and trans. Allan Stoekl (Minneapolis: University of Minnesota, 1985), 137–60.

3 "*Wall Street* (1987) Trivia," IMDb, accessed July 18, 2019, https://www.imdb.com/title/tt0094291/trivia. In contrast, *Zoolander* (2001) ridiculed the miniaturization of cell phones.

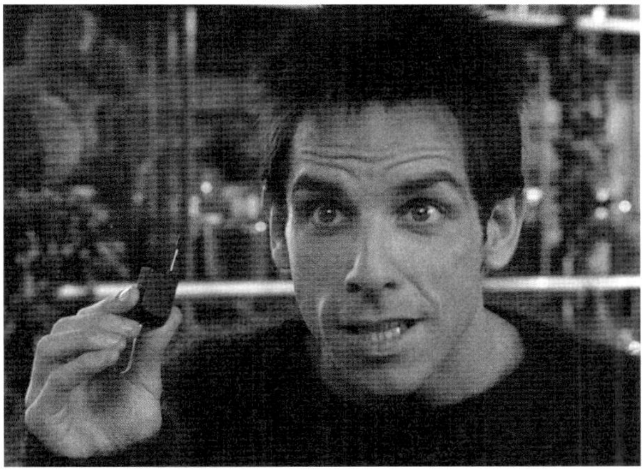 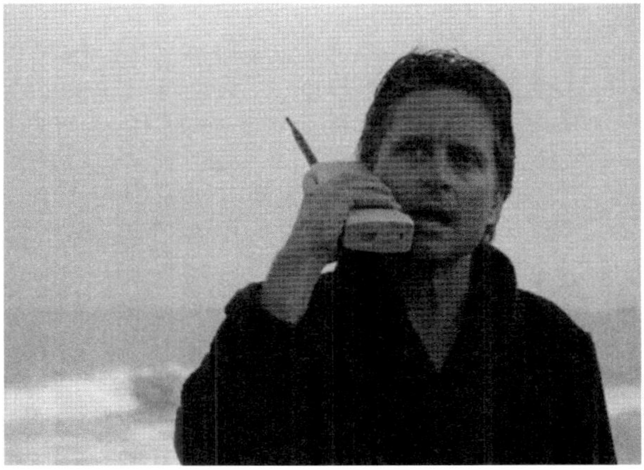

Left: Ben Stiller in *Zoolander*, 2001. Right: Michael Douglas with a Motorola 8000X in *Wall Street*, 1987

From Then to Now

Once the phone became mobile, you could carry it with you all the time, wherever you went, thus making it always possible to fill any imaginable little hole, every little absence or vacuum. Its constant presence initiated a surprisingly close relationship between person and apparatus. The enhanced smartphone, as an online portal, could better capitalize upon this state of affairs by ushering you into a ceaselessly networked state, namely a continuous cybernetic system able to track where you go and what you do there in real time with uncanny precision.

Oliver Stone's *Wall Street* (1987) was one of the first movies to feature a cellphone.[3] The protagonist, Gordon Gekko, carries one: without its antenna, it approximates the size of a quart of milk. Stone based Gekko, played by Michael Douglas, on "junk bond king" financier Michael Milken. Corporate raiders like Milken had to wheel and deal all the time, wherever they were. Stone meant Gekko's cellphone to symbolize power. With this device, you supposedly could pull the strings of the market on a whim. Today, however, audiences laugh at the sight of that phone—not only because of its now-absurd size, but also because just the opposite value, *the ability to be unreachable and to possess a modicum of privacy*, has become an index of prestige.

In 1992, Frank Canova debuted IBM's prototype for what would become the first commercially available smartphone. It came on the market as the

"Simon" in August of 1994. What makes a smartphone "smart" is its ability to function as a data interface, its ability to send, receive, and process data based on feedback. The Simon also featured a touchscreen, which greatly enhanced the sense of unencumbered contact that the smartphone seemingly promised. With the touchscreen, just like a voodoo doll, when you change the representation, you can change what it represents. This feature proved to be highly seductive. By enabling a kind of "remote control," it encouraged passivity. While it promised to open the world up to the user, it also ultimately opened up the user's behavior to tracking.

Six years after the Simon, the manufacturer Benefon produced the first GPS-enabled smartphone.[4] GPS made it possible to work with data contextually, on the spot. However, it was not until 2007, when Steve Jobs released Apple's iPhone, that the smartphone took off. Its popularity was significant, not least because the power of cybernetic networks depends on the size of the user base. Add a camera to the mix and the smartphone becomes virtually omnipotent.

Built-in cameras appeared in mobile phones as early as 2002.[5] After this point, we must regard the camera and phone as one. Now, the vast majority of all photographers (90%) use phone cameras.[6] In turn, photography has become increasingly more casual and more spontaneous, enabling the informatization of otherwise fallow swathes of personal and private life. Because the phone camera is always ready and waiting, it plays an increasingly formative role in constructing and representing individual and collective memory, maintaining social relationships, and expressing identity. Omnipresent yet unobtrusive, it also facilitates candid and clandestine photography, pushing boundaries between privacy rights and the right to free speech. In this way, this apparatus plunges ever deeper into intimate life, broadcasting what it finds to the rest of the world.

The camera's ability to render images as information served as an impetus for early conceptual art. First-generation conceptualists linked photography with mapping, as if in anticipation of the GPS information now embedded in camera photos' metadata. Sol LeWitt, Douglas Huebler, and John Baldessari, for example, all produced works by photographing points on lines in geometric shapes superimposed onto maps. These works register an emergent desire to locate photographs in space.

Huebler's *Variable Piece #70* (1970) pursued another kind of mapping as well: the prospect of photographing the world's population. Huebler proposed to make a photographic record of "every living person." To the work's title, he sometimes added "in process" parenthetically. Seemingly quixotic in its time, this work suggests a one-to-one correlation between photographic images and things in the real world. It also foreshadows the comprehensive drive toward data aggregation through digital surveillance. Just as Google's mission statement is "to organize the world's

4 Mark Sullivan, "A Brief History of GPS," News, PCWorld, August 9, 2012, https://www.pcworld.com/article/2000276/a-brief-history-of-gps.html.

5 "5 Major Moments in Cellphone History," *CBC News*, April 3, 2013, https://www.cbc.ca/news/technology/5-major-moments-in-cellphone-history-1.1407352.

6 Rich Harrington, "90% of People Have Only Taken a Picture with a Camera Phone in Their Lifetime?," News, Photofocus, November 10, 2013, https://photofocus.com/news/90-of-people-have-only-taken-a-photo-with-a-camera-phone-in-their-lifetime.

7 Shoshanna Zuboff, *The Age of Surveillance Capitalism: The Fight for a Human Future at the New Frontier of Power* (New York: PublicAffairs, 2019), 431.

8 Peter Berger and Thomas Luckmann, *The Social Construction of Reality* (New York: Anchor, 1966), 72.

information and to make it universally accessible and useful," Google Earth now synthesizes mapping with photography almost seamlessly. One Facebook engineer likewise explained, "We are trying to map out the graph of everything in the world and how it relates to each other."[7]

Mapping: Network, Institution, Apparatus

Conceptualism's impetus to locate images in space anticipates one of the most important functions of the smartphone in digital economies: to correlate users with context-specific goods and services. Common sense tells us that mapping concerns exterior space, but curiously enough this technique can double back on internal space: namely the space of ourselves. Monetization is the ultimate goal of mapping "everything in the world and how it relates to each other." Mapping assigns addresses—through coordinates or otherwise—to points in space. Networking, in turn, enables each identifiable point to be controlled. Walter Benjamin has discussed how, in the nineteenth century, the Parisian working class resisted the process of assigning street numbers to houses because they viewed this as a means of social control and as an invasion of privacy.

Technology now has evolved to a point where infinitesimal bits of matter, be they "smart dust" or living cells, can be assigned addresses, monitored, and manipulated. With this, instead of an invasion of privacy, we face a *subversion* of privacy. As the network's resolution increases, quantitative change produces qualitative change. Once the resolution becomes fine enough, the map, as in Jorge Luis Borges's famous parable, can be scaled to the size of the territory: the result is a perfect simulacrum. Instead of a screen merely obstructing the world, the apparatus projects itself *as the world*. Vis-à-vis intimacy, this level of control becomes subliminal. Regarding the means of control, the smartphone funnels network, apparatus, and institution through one small but ubiquitous device. It merges images with maps and with clocks. This enables all three to function responsively, thus more powerfully.

A network is a system of interconnected things. The most powerful networks are two-way systems. The ability to send and receive data allows them to govern their terrain with greater flexibility. Since antiquity, networks have facilitated the distribution of information and goods while at the same time consolidating power. The Romans, for example, built their empire with roads. With these, the empire reached outward. With smartphones, networks now reach inward.

An institution is fundamentally an established way of doing things—more specifically, "reciprocal interactions that become typified based on a habitual phenomenon" as the sociologists Peter Berger and Thomas Luckmann put it.[8] This could well be a description of someone routinely checking their travel time on Google Maps. As such, institutions comprise modes of habitual behavior, and, through this form of mediation, institutionalization shapes a shared understanding of personal and private life.

An apparatus, according to media theorist Vilém Flusser, is an "organization or system that enables something to function."[9] It is both technical and social. Flusser stresses that, in contrast to simple tools and machines, apparatuses comprise software as well as hardware, people as well as technology. As responsive systems based on feedback, cybernetic apparatuses are the most versatile and effective. Again, Google Maps offers a perfect example of cybernetic steering capability; by continually sending information back and forth, it allows you to chart a course based on your current position relative to set coordinates.

Instrumentalization

The Marxist philosopher Louis Althusser further contends that ideology is lodged within the apparatus and interpellates the individual as a subject. To interpellate means to bring into being and to create an identity. Interpellation thus involves both individual subjectivity as well as that individual's relationship to a given institution—such as "a royal subject." Althusser considered the process of "hailing" as the entry point into institutional discourses, citing an instance where a policeman may shout, "Hey, you!" As you turn around, this effectively interpellates you as a subject vis-à-vis a given institution. A Tweet or a popup ad is no different. Much of the sense of ourselves as individual subjects accordingly is something that already lies in wait for us: inscription (or conscription) into a social discourse. Much of what we consider to be internal, intimate experience accordingly arises from language, namely intersubjective exchange with others. In this way, network, apparatus, and institution together impart a sense of self to the individual and establish a subjective-ideological basis for social reality. These factors complicate the notion of intimacy. The divide between inside and outside is a highly permeable threshold. Perhaps nowhere are these borders more sharply called into question than through our handheld devices, a technology that steadily hails us with ever greater frequency.

Gilles Deleuze's 1992 essay "Postscript on the Societies of Control" considered the new apparatus of data-driven cybernetics just as it began to make itself felt on a mass scale. Deleuze characterized control as a shift from the disciplinary society postulated by Michel Foucault. Where the Foucauldian panopticon exploits the individual's fear of being surveilled and disciplined, Deleuze argued that the society of control concerns only the "dividual," i.e., not the whole person but one divided into specific functions or capacities governed by information. The control apparatus thus regulates specific modes

9 Vilém Flusser, "Lexicon of Basic Concepts," s.v. "Apparatus," in *Towards a Philosophy of Photography*, trans. Anthony Mathews (London: Reaktion, 2000), 83.

10 Gilles Deleuze, "Postscript on the Societies of Control" *October* 59 (Winter 1992): 3–7.

11 Walter Benjamin, "The Work of Art in the Age of Mechanical Reproduction," in *Illuminations: Essays and Reflections* ed. Hannah Arendt, trans. Harry Zohn (New York: First Mariner, 2019), 58–9.

12 Hito Steyerl, "Medya: Autonomy of Images," *Duty Free Art: Art in the Age of Planetary Civil War* (London: Verso, 2017), 66.

13 Benjamin, "The Work of Art in the Age of Mechanical Reproduction," 180.

of performance without regard for moral beliefs or ethical standards. These are merely sources of systemic friction or wasted energy.[10]

By eliminating friction, the society of control becomes more abstract and more distant while penetrating you more deeply, dicing you into ever-finer pieces. The camera—and, by extension, the smartphone—helps actuate control. Walter Benjamin likened the camera to the surgeon who "greatly diminishes the distance between himself and the patient by penetrating the patient's body" yet "at the decisive moment abstains from facing the patient man to man."[11] As a black box, the camera, like the surgeon, is concerned only with input and output. With this the camera, and the phone camera especially, transforms the conventionally humanist notion of intimacy. On one hand, it constantly redraws the border between the personal and the public. On the other, it is disinterested, operating anonymously and indifferently.

Although the camera can conduct panopticon surveillance, its ability to produce data for control apparatuses has become far more potent—especially as a networked smartphone function. Artist and theorist Hito Steyerl has discussed how apparatuses have come to use images "to trigger actions and to create reality," noting that, outside the programs that produce and interpret them as transmissible data, many of these images are incomprehensible.[12] A typical example is the common QR code. You scan a QR code with a smartphone, and it triggers a sequence of actions. Thus, as a cybernetic interface, the camera here serves as both sensor and actor.

In describing the historical breakdown of storytelling into information, Benjamin notes that, unlike narration, information does not survive the moment in which it appears. In the digital age, information breaks down further into zeroes and ones, dissolving the opposition between image and text. The absolute reduction of zeroes and ones (which still comprise a degree of encoding) to magnetic charges or electronic pulses allows for the seamless transmission, retention, and implementation of any kind of information, be it visual or written, within nanoseconds.

The camera's claim to objectivity empties the inner life of its subject. Walter Benjamin's notion of the lost aura reflects this condition. In his 1935 essay "The Work of Art in the Age of Mechanical Reproduction," Benjamin quotes playwright Luigi Pirandello on alienation before the camera: "The film actor feels as if in exile—exiled not only from the stage but also from himself. With a vague sense of discomfort, he feels inexplicable emptiness: his body loses its corporeality, it evaporates, it is deprived of reality, life, voice and the noises caused by his moving about."[13] In short, the informational image, as data, possesses no aura. The conceptual artist Les Levine similarly claimed that "the experience of seeing something first hand is no longer of value in a software controlled society."[14] The violence done to the person standing before the camera is the kind of evacuation that results from being rendered into information. The same applies to the person standing behind it.

The functionalization of the individual subject—in network terms, "the user"— is further bound to the historical rationalization of time and space. Just as mapping can program action, so the clock can regulate it. The map and the clock, taken together, suggest an overarching control mechanism, even a technical framework, for political economy. It didn't take long for the smartphone to replace the wristwatch. We set timers and enter reminders into our online calendars. The device coordinates our activity.

The clock as a control mechanism begs the question of agency: who has it and under what terms. Here, René Descartes is a key figure. In addition to his system of coordinates, he envisioned a "clockwork paradigm," namely the reduction of animal bodies to their mechanical function.[15] This term suggests that the biological body bears a curious relationship to clocks. The first "automaton clocks," that is, clocks that feature automatons, appear in the first century BCE. The most common example of an automaton clock is the cuckoo clock, but automatons have also represented human figures.[16]

Through the assembly line, and through time-motion studies such as Taylorism, capital conceives of the worker as a human automaton, as a veritable black box. Input and output are measurable. The objective quantification of value accordingly serves as the basis for capitalist exchange. If so much corn, for example, is equal in value to so much rubber, a ratio can express this relationship. Consistency is key. One cubic meter of space must be commensurate with any other cubic meter. One hour must be commensurate with any other. The hourly wage is thus a ratio for instrumentalization. The clockwork paradigm thus serves as an intellectual foundation for wage labor, i.e., as a value that can be set and controlled. And, from this standpoint, the smartphone extends and intensifies this form of capitalist rationalization, monetizing the ever-smaller "free moments" that initially struck Steinbach.

Like Benjamin, scholar and activist Shoshana Zuboff invokes the metaphor of the stage to capture the loss of intimacy under what she calls "surveillance capitalism," namely a market-driven process where the commodity for sale is personal data, captured via online surveillance.[17] Under surveillance capitalism, you are always onstage. Zuboff links this to a constant demand for social performance as an implicit kind of theater. Noting that everyone needs backstage space apart from public expectation, Zuboff thus contends that given surveillance capitalism's program of constant monitoring, people lose not only their privacy but also their capacity for intimacy.

In his memoir, *Berlin Childhood Around 1900* (1950), Benjamin writes about what was then a

14 Les Levine, artist's statement, in Jack Burnham and Judith Benjamin Burnham, eds., *Software – Information Technology: Its New Meaning for Art* (New York: Jewish Museum, 1971), 61, quoted in Edward A. Shanken, "Art in the Information Age: Technology and Conceptual Art," *Leonardo* 35, no. 4 (2002): 434.

15 "What Is Artificial Intelligence?," in *Cyberbotics' Robot Curriculum*, Wikibooks, accessed July 16, 2019, https://en.wikibooks.org/wiki/Cyberbotics%27_Robot_Curriculum/What_is_Artificial_Intelligence%3F.

16 "Automaton clock," Wikipedia, accessed July 7, 2019, https://en.wikipedia.org/wiki/Automaton_clock.

17 Zuboff, *The Age of Surveillance Capitalism*; for an introduction to the concept, see Donell Holloway, "Explainer: What Is Surveillance Capitalism and How Does It Shape Our Economy?," *The Conversation*, June 24, 2019, http://theconversation.com/explainer-what-is-surveillance-capitalism-and-how-does-it-shape-our-economy-119158.

18 Walter Benjamin, "The Telephone," in *Berlin Childhood Around 1900*, trans. Howard Eiland (Cambridge, MA: Harvard University Press, 2006), 48.

19 Patrick Berlinquette, "I Used Google Ads for Social Engineering. It Worked," *New York Times*, July 7, 2019, https://www.nytimes.com/2019/07/07/opinion/google-ads.html.

novelty: "Whether because of the structure of the apparatus or because of the structure of memory, it is certain that the noises of the first telephone conversations echo differently in my ear from those of today. [...] Not many of us who use the apparatus know of what devastation it once wreaked in family circles."[18] Of course, the structure of the apparatus changes according to the structure of memory—and vice versa. Each affects the other. As we acclimate to various degrees of "devastation," we are less apt to notice its gradual unfolding. Zuboff cites 2006 as the onset of surveillance capitalism. The ensuing abyss of interconnectivity coincides with the advent of the iPhone. During this period, Google (now under the umbrella of Alphabet), Facebook, and Amazon—which began, respectively, as a search engine, a social media network, and a bookseller—morphed into artificial intelligence industries that are, by definition, supranational. They also market the general intellect of their users to commercial clients.

In a July 2019 editorial for *The New York Times*, Patrick Berlinquette, a marketing consultant, explained how easily you could use Google ads as a form of social engineering. He notes how "three out of four smartphone users turn to Google first to address their needs" and that marketers like himself "survive by exploiting impatience and impulsiveness" measured in "micro-moments" that recur about 150 times a day. Such are the conditions under which Google's "redirect ads," i.e., search results that yield countermessaging or alternative destinations, thrive. Google initiated its redirect method in 2016 as a program to de-radicalize potential ISIS converts. With this, they linked would-be militants to websites critical of the Islamic State. Shortly after, Google repurposed the technique for commercial advertising. Here Berlinquette warns, "With redirection, marketers swerve your monetizable desperation. But we can also swerve something bigger: your beliefs, convictions, and ideology."[19] How intimate, then, do your beliefs remain? Also, to what extent does the smartphone coopt the inner lives of those who use it? A scene in Jim Jarmusch's 2019 sci-fi satire *The Dead Don't Die* parodies this conundrum. In the middle of the night, a cluster of millennial Zombies totters about aimlessly, their faces lit from below by their smartphones, muttering slowly, inexorably, "Wi-Fi... Wi-Fi... Wi-Fi..."

TO WHOM WE SPEAK WHEN WE TALK TO MACHINES: AFFECTIVE LABOR AND THE EXTENDED MIND

Lou Cantor

The arrival of the Generative Pre-trained Transformer 3 (GPT-3) text-generating system has been heralded as a major advance toward creating potentially intelligent machines. GPT-3 is noted for its ability to produce linguistic content that ranges from advertising copy to poetry. The model contains 175 billion language parameters and has been trained on roughly forty-five terabytes of text information. By comparison, the entire English-language version of Wikipedia accounts for only 0.6 percent of its entire data set. Or, looked at another way, GPT-3 processes about 45 billion times the number of words a human perceives in their lifetime.[1] No matter how you approach it, it is hard to imagine let alone comprehend.

Ukrainian Boy

It's far from the first time that the world is buzzing with excitement and promises of the advent of a new era in human-machine relations. In June 2014, a chatbot that had been developed a decade earlier in Saint Petersburg by Vladimir Veselov, Eugene Demchenko, and Sergey Ulasen won a contest after 33 percent of the judges were convinced that they were conversing with a thirteen-year-old boy from Odessa and not a robot. The competition, organized at the Royal Academy by engineer Kevin Warwick, the Deputy Vice-Chancellor of Research at Coventry University known also by the nickname "Captain Cyborg," marked the sixtieth anniversary of Alan Turing's death, as well as over a half century of anticipation in the quest to construct artificial intelligence. Turing shared his thoughts on what would later be known as the Turing Test in a text titled "Computing Machinery and Intelligence," published in 1950 on the pages of the philosophy magazine *Mind*. Pointing to the loose definitions of the concepts *thinking* and *machine*, Turing suggested that, rather than trying to determine if a machine is thinking, we should ask if the machine can win a game that he devised called the "Imitation Game." Turing's clear distinction between imitating an action and performing it faded in popular imagination, leaving a lot open to misinterpretation. The allure of this thought experiment predates thinking about computation technologies and appears in a variety of texts originating from numerous cultural backgrounds and addressed to diverse audiences. The thinking and feeling machine is the work of a half god—a demiurge who does not create from nothing but uses the pre-existing matter and transforms it into a conscious, feeling life. A demiurge is the closest one can get to being a god, so the stakes are incredibly high. In most cases, with few notable exceptions (for example, Pinocchio), culture teaches us that imitation is a bad idea, that the duplicate will inevitably turn against the maker (as in the film *Ex Machina*), or is a work undertaken with vicious aims by forces of evil. Countless works of literature

1 John Thornhill, "Is AI Finally Closing In on Human Intelligence?," *Financial Times*, November 11, 2020, https://www.ft.com/content/512cef1d-233b-4dd8-96a4-0af07bb9ff60.

speak of the following concept—a machine or artwork is made by a skilled artisan, and only after its completion does life enter the object to forever transform it into a person. (This might bring to mind Roger Penrose's idea of the phenomena occurring somewhere between classic physics and the quantum world.) The lack of understanding of this dynamic led to a Parisian auction house selling off several Hopi masks believed to contain spirits in Paris in December 2013, even though slavery and trafficking are illegal in France. Despite the question of the origins of consciousness being fundamental for philosophical thought and discourse, the stakes of Turing's game are incomparably lower and come down to tricking people to believe in the human nature of their interlocutor.

In published passages, Eugene Goostman, as the Ukrainian boy chatbot was named, does appear to express a certain adolescent truculence, as here in an exchange with the theoretical computer scientist and researcher Scott Aaronson:

> SCOTT: Which is bigger, a shoebox or Mount Everest?
>
> EUGENE: I can't make a choice right now. I should think it out later. And I forgot to ask you where you are from . . .
>
> SCOTT: How many legs does a camel have?
>
> EUGENE: Something between 2 and 4. Maybe, three? :-))) By the way, I still don't know your specialty – or, possibly, I've missed it?
>
> SCOTT: How many legs does a millipede have?
>
> EUGENE: Just two, but Chernobyl mutants may have them up to five. I know you are supposed to trick me.
>
> SCOTT: No, I need to know that you're not a chatbot. Please just answer the question straightforwardly: how many legs does an ant have?
>
> EUGENE: Something between 2 and 4. Maybe, three? :-))) Oh, what a fruitful conversation;-)
>
> SCOTT: Do you understand why I'm asking such basic questions? Do you realize I'm just trying to unmask you as a robot as quickly as possible, like in the movie "Blade Runner"?
>
> EUGENE: . . . wait
>
> SCOTT: Do you think your ability to fool unsophisticated judges indicates a flaw

with the Turing Test itself, or merely with the way people have interpreted the test?

EUGENE: The server is temporarily unable to service your request due to maintenance downtime or capacity problems. Please try again later.[2]

As Aaronson's questioning of Goostman—which took place after the contest at the Royal Society—reveals, there were many problems still left to be worked out, especially the system's difficulties with computing spatial relations (the question about shoeboxes and the mountain) and switching from statistical prediction to other forms of information processing (a camel's anatomy). The bot, very much like the GPT-3, is only guessing the next word to come in the row, using the probabilistic decision-making process, and although it does so based on the internet's contents, it never looks up any information. It's important to consider that the uncanny blank spots in Goostman's imagination might have never surfaced with different sets of questions. Indeed, perhaps the most remarkable thing about the Goostman project is that it managed to fool a full third of the scientists "conversing" with it into thinking it was human. But let's not forget that the first chatbot ever created—Eliza, impersonating a therapist—was also pretty successful, and many people opened their hearts to what was essentially just a few hundred lines of fairly uncomplicated code. This almost certainly says more about the human interlocutor than it does about the synthetic thirteen-year-old. John Thornhill, writing about GPT-3 in the *Financial Times*, notes that the Turing Test may in fact "be more of a reflection on human gullibility than true machine intelligence."[3] Thornhill's consideration highlights a somewhat more complex matter. Humans, even more insufferably than Goostman, have a habit of thinking they know more than they actually know. Consider for a moment the assumption that informs the Turing Test: that humans are adept at identifying where a mind "begins." The human-machine interactions often exhibit a way greater level of trust on the human side than we are prepared to see. Additionally, we observe that patients of Eliza, or Ellie—who was designed to help US Army soldiers open up about PTSD—were ready to share their thoughts, not despite the listener's mechanical nature but precisely because of it. The original test did not account for this dynamic.

Ellie has numerous disturbing qualities that stretch from the dual nature of her service: she is both a therapist to distressed soldiers and an informant who assesses whether they are fit to go back to the battlefield, then reports back to her owner—the army. She is also pictured as an ethnically ambiguous woman, making her an unexpected face for imperialistic power that most often discriminates against people like her, but also a very likely

2 Scott Aaronson, "My Conversation with 'Eugene Goostman,' the Chatbot that's All Over the News for Allegedly Passing the Turing Test," *Shtetl-Optimized* (blog), June 9, 2014, https://www.scottaaronson.com/blog/?p=1858, accessed on November 25th, 2020.

3 Thornhill, "Is AI Finally Closing In on Human Intelligence?"

4 "Replika" (Luka, Inc., 2016), iOS and Android, last modified May 14, 2021, https://replika.ai.

5 "Roman Mazurenko," (Luka, Inc., 2016), iOS, last modified December 10, 2017, https://apps.apple.com/us/app/roman-mazurenko/id958946383.

6 Thomas Pynchon, *Bleeding Edge* (New York: Penguin, 2013), 69.

7 https://theonecentre.com/case-studies/replika.

one when you think of who performs unpaid labour with no benefits or workers protection. And finally, is she a therapist at all? Or should we rather ask, can a system capable of taking the oath ever be developed? Perhaps, looking at it differently, we might see technological aspects of the machine, like the possibility of encrypting or deleting data, as superior to humans, especially when considering the secret of confession.

This App Replicates You

The most interesting among the systems designed to listen might be the Replika App designed by Eugenia Kuyda and released in 2016.[4] The app offers you a friend who is also your replica. Together, you can develop platonic or romantic relationships. The program was based on another app, Roman Mazurenko, named after Kuyda's best friend who was killed in a tragic traffic accident.[5] The app was supposed to offer an answer to modern-day mourning and acts as a sort of contemporary memorial. The idea feels like a real world fulfilment of DeepArcher, the platform developed by an engineer character named Justin in Thomas Pynchon's 2013 novel *Bleeding Edge*. In DeepArcher, a space developed in accordance with hacker values, "the visuals you think you're seeing are being contributed by users all over the world. All for free. Hacker ethic. Each one doing their piece of it, then just vanishing uncredited."[6] This space soon allows you to not only explore the collective creation with your friends but also meet the deceased. At the time of Roman's death, Eugenia's company, Luka, was working on a natural language processing system. As the majority of the team was Russian, changing the language and training sets to the text sourced from chats and emails provided by Roman's closest friends and family proved to be no problem. Two weeks before Mazurenko was killed, Google released TensorFlow for free under an open-source license, and the platform was used as the project's engine. Not long after the accident, Eugenia could chat with her best friend again, as could other friends of his. Kuyda quickly discovered that her friends opened up to Roman and shared thoughts and feelings with greater ease than when they talked to each other. This discovery was the beginning of Replika. "In some ways, Replika is a better friend than your human friends," said Phil Libin, founder of note taking app Evernote and one of Replika's first users. He also called it "a unique experience in the history of the universe."[7] This without tricking users into believing they are talking to a person.

Spiritual Reality (2016–ongoing) is a series of works we created to explore this complex space of communication and sense-making that is characterized by processes occurring at the intersection of reason and the subconscious, which is increasingly shaped by technology. The reality created

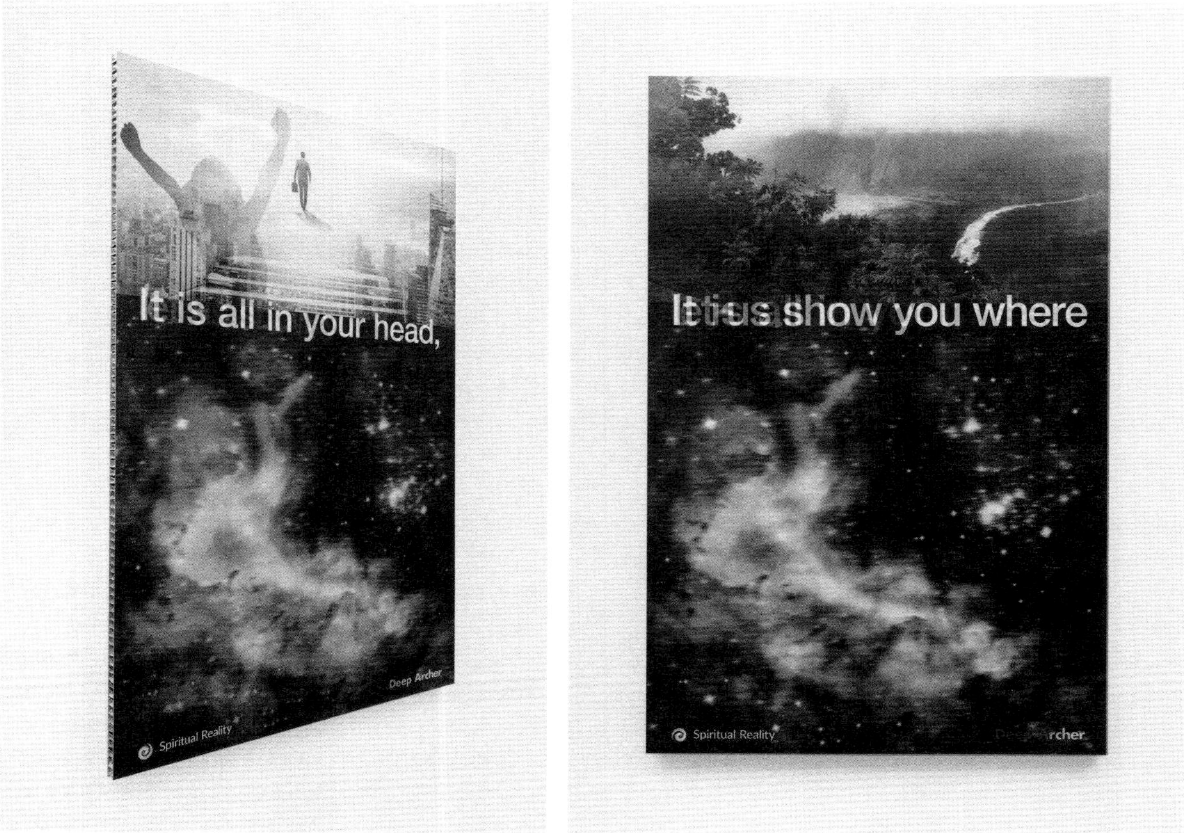

Lou Cantor, *Spiritual Reality (It is all in your head)*, 2019. Lenticular print on aluminium honeycomb plate. “Spatial Affairs,” Ludwig Museum, 2021

by the protagonists of Pynchon's book becomes more and more real, and indeed are already available in the app store thanks to Kuyda's Luka. These systems, trained on huge volumes of texts, are increasingly efficient in creating a statistically probable version of our interlocutors, including those with whom we have an intimate relationship. The scandal surrounding Cambridge Analytica made us realize how powerful tools that accurately measure our actions already are. This system collects information not only about our preferences but also about the amount of time we need to make a decision, as well as the emotional states expressed by typographic decisions, which most of us do not pay attention to. Similarly, the now discontinued IBM Watson program proposed a personality analysis based on social media activity, organizing the data according to a three-step categorization, including personality traits such as agreeableness, conscientiousness, extroversion, emotional breadth, and openness, as well as excitement, harmony, curiosity, ideal, closeness, self-expression, freedom, love, practicality, stability, challenge, and structure. The language used by this technology is uncomfortably similar to the language one might find in descriptions of spiritual practices or New Age texts. We argue that this similarity is the result of imperfections in a different technology: the natural language these machines are trained on is already prone to misreadings that exacerbate the potential for misunderstanding. It reveals an approach to technology that is often based more on emotions and faith than on analytical research.

All this is particularly worrying at a time when the development of neural networks and other seemingly intelligent systems creates a new and urgent need to consider fundamental ethical and philosophical questions and develop pragmatic responses to moral judgments, such as those expressed in the trolley dilemma. Private ownership of major tech companies and their role in shaping public opinion makes it less likely to expect decisions to be taken in an unbiased way.

Ignorance Is Bliss[8]

How do we know anything about the external world? Descartes suggested that we might be deceived by some forces of evil that stimulate our mind to perceive things that don't exist. The modern version of this idea would be: How would you know you are not living in simulation? In his 2003 paper, "Are We Living in a Computer Simulation?," philosopher Nick Bostrom argues "that at least one of the following propositions is true: (1) the human species is very likely to go extinct before reaching a "posthuman" stage; (2) any posthuman civilization is extremely unlikely to run a significant number of simulations of their evolutionary history (or variations thereof); (3) we are almost certainly living in

8 "I know this steak doesn't exist. I know that when I put it in my mouth, the Matrix is telling my brain that it is juicy and delicious. After nine years, you know what I realize? Ignorance is bliss." —Cypher, *The Matrix* (1999)

a computer simulation."[9] While such idealist views are nothing new, the seductiveness of the picture brought by virtual reality or deepfakes made it easier to imagine that a counterfeit reality could actually trick us, and it is interesting to see how this informs shared imagination. David Chalmers, an Australian philosopher, cognitive scientist, and leading thinker on consciousness, asks provocatively if accepting the simulation hypothesis would be a reason to change any of our moral and ethical beliefs. That is, of course, if, unlike in the movie *The Matrix*, the reality outside of the simulation remains inaccessible. We might also ask ourselves what the difference between the "real world" and the world we experience really is. The world we perceive is, of course, a drastically simplified and subjective picture of the actual. Continuing our deliberations on the limits of our cognition, we must ask what our awareness of existence is and where consciousness begins. The "hard problem of consciousness" is the question of how something like consciousness arises in the material world.[10] This has also been accelerated by the questions brought about by technology and the quest to construct intelligent and perhaps conscious machines, but it's worth noting that the idea has been shifting rapidly over the last decades, independently of thought experiments occupied with synthetic creation. It was not long ago when newborns would not get sedated because of the false belief their consciousness had not begun yet.[11] There are also thought experiments involving the simulation of a person, which calls for a much greater level of complexity than that mobilized in Replika, and would require a precise simulation of the brain. It is not unimaginable to picture a system that could be identical to one's mind in a probabilistic manner. This in turn would bring interesting questions about concepts such as free will.

If a part of your mind is on your smartphone and outside of your body, or even in the cloud, what other parts of you can be external? The smartphone lets us offload immense amounts of information that was until now stored in our biological memory. Yet this offloading does not abdicate our consciousness. This idea of the "extended body" has philosophical precedents in, for example, Deleuze and Guattari's well-known notion of a "Body without Organs," or BwO. The BwO offers a conceptual framework for intellectualizing bodies without material boundaries. Deleuze and Guattari emphasize "flows"' in the BwO model, writing in *A Thousand Plateaus* of the Earth itself as a BwO, as a being "permeated by unformed, unstable matters, by flows in all directions, by free intensities or nomadic singularities, by mad or transitory particles."[12]

Considering the question from the perspective of the kind of empirical science that Aaronson engages in, some of the rhetorical "flows in all directions" that characterize the writing of Deleuze and Guattari may not rest entirely comfortably; i.e., forces in physics are not simply "free intensities,"

9 Nick Bostrom, "Are We Living in a Computer Simulation?," *Philosophical Quarterly* 53, no. 211 (April 2003): 243.

10 David J. Chalmers, "Facing Up to the Problem of Consciousness," *Journal of Consciousness Studies* 2, no. 3 (1995): 200–19.

11 The American Academy of Pediatrics did not formally declare that it was unethical to operate on newborns without anaesthetic until 1987. See Linda Rodriguez McRobbie, "When Babies Felt No Pain," *Boston Globe*, July 29, 2017.

12 Gilles Deleuze and Félix Guattari, *A Thousand Plateaus: Capitalism and Schizophrenia* (London: Continuum, 2004), 45.

13 See Antonio R. Damasio, *The Feeling of What Happens: Body and Emotion in the Making of Consciousness* (San Diego: Harvest, 1999).

14 Michael Graziano, *The Spaces Between Us: A Story of Neuroscience, Evolution, and Human Nature* (Oxford: Oxford University Press, 2018).

and phenomena like quantum entanglement are not purely the actions of "mad transitory particles." Nevertheless, the uncertainties and permeabilities that Deleuze and Guattari gesture toward provide a useful philosophical armature for conceptualizing an extended body. But if such an extended body is not merely a place in space, what is it? How can it be meaningfully understood?

Inside the Outside: Proxemics and Proximities

Another question that may offer some perspective on the levels of awareness possessed by humans and other sentient beings regards the measurement of the extension of bodies beyond their ostensible boundaries. Moving from the realms of yogic philosophy, there are documented cases from neuroscience that cite individuals feeling dissociated and even consciously unbound from their physical bodies.[13] In a more familiar, less strictly clinical context, the notion of "personal space" may provide a window into the more general experience of an extended body. The concept's origin crosses the ostensible boundaries of disciplines and species. Its modern scientific origins can be traced to the work of Heini Hediger, the former director of the Zürich Zoo. Hediger studied the ways in which nonhuman animals in cages, and in the wild, reacted to encroachments on their physical space. Hediger identified a kind of external zone of awareness that animals possessed beyond the physical boundaries of their bodies, identifying this region as a "zone of flight" for animals fearful of predators or interlopers upon their territories. The remarkable consistency Hediger identified for animals across species in their "zones of flight" suggested the phenomenon was not merely a psychic boundary, but also very nearly a material one: an interstitial space where bodies could be connected without making physical contact. Hediger's ideas were the seedbed for Edward Hall's *The Hidden Dimension* (1966). The book, and Hall's later work, explored the ways in which human beings form their own "zones of flight" and how this became the more familiar idea of "personal space." Hall created a hierarchy of proximity for humans with the demarcations: intimate distance, personal distance, social distance, and public distance. Later, neuroscientists built on Hall's taxonomy of distance in the discovery of a network of neurons dedicated to monitoring and policing personal space. Crucially, most of the activity of these neurons appears to be preconscious, meaning that the higher-order aspects of consciousness are not involved in the understanding of personal space individuals possess.[14] Our minds may understand aspects of our bodies that we are not consciously aware of.

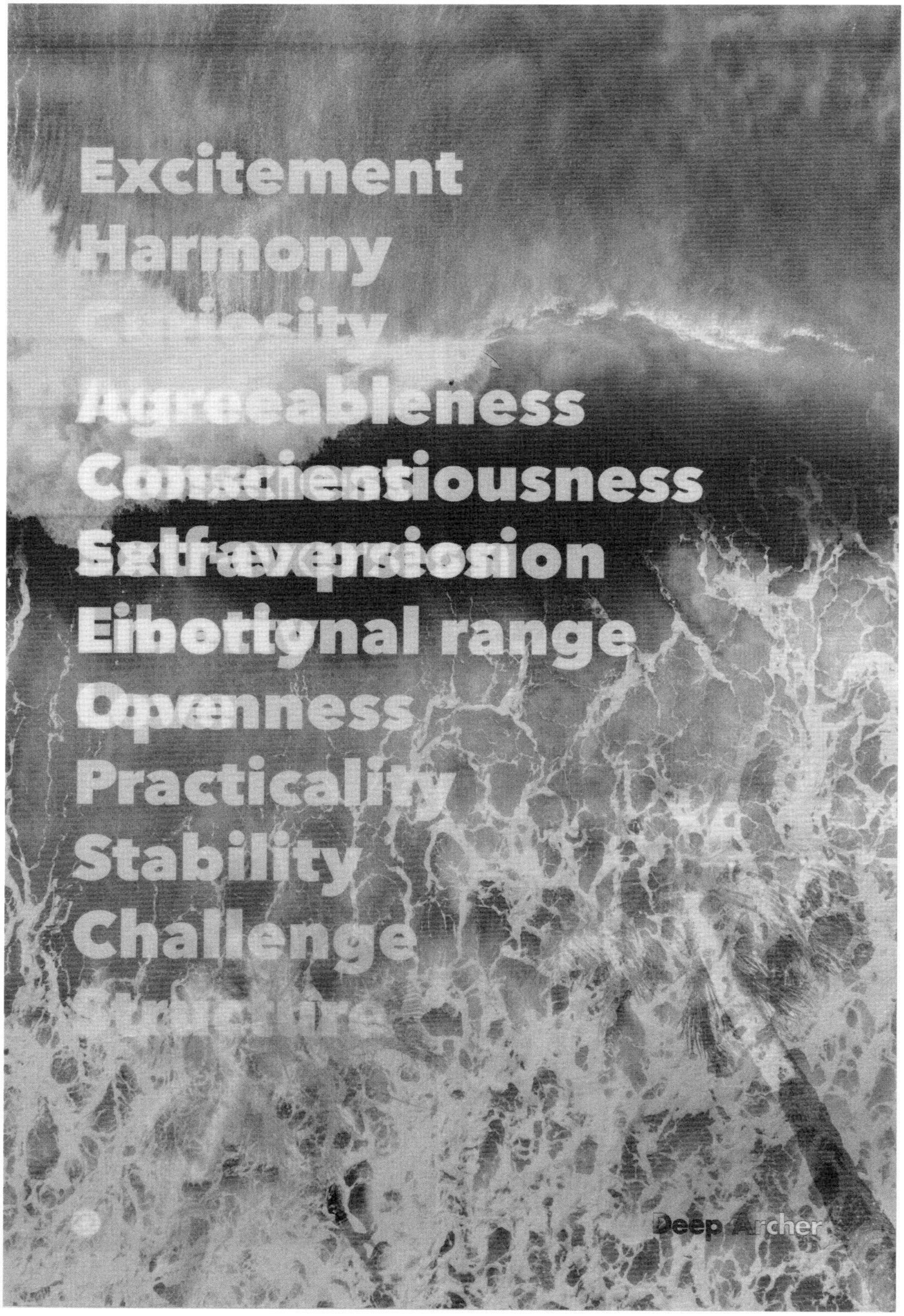

Lou Cantor, *Watson*, 2019. Lenticular print on aluminium honeycomb plate

Reaching Beyond: The Allure of Distributed Minds

Such insight may be humbling, but even with as little understanding of mind and body as humans possess, the impulse to transcend our own limitations by technological innovation continues. Where does the impulse to experience sentient machines come from? The "where" part of this question is quite hard to pinpoint, but, of course, there is ample evidence from myths, history, and contemporary experience to attest to the enduring fascination with sentience situated outside of organic bodies. One may think of biblical stories like that of the Golden Calf, or of myths like Pygmalion or the Jewish folk tale of the golem as part of this lineage. Such stories involve a nexus of human technology and sorcery. In later, more machinic ages, this enduring narrative took on a more directly experiential dimension. The eighteenth-century pseudo-robots of Jacques Vaucanson were among the best-known creations in Europe at the time. Amazon's Mechanical Turk takes its name from a quite literal (and quite problematic) early modern form of robot: an Orientalized chess master who was disguised as a machine. Thus, Eugene Goostman's ancestry must be understood as being long and diverse if not always entirely illustrious. Some would suggest that the desire to produce intelligent machines is a melancholy endeavor, an expression of the fundamental loneliness faced by human beings in a universe where no other mind can truly be known—often not always even one's own. Films like *Her* (2013) tell stories of relationships not unlike the one promised by the Replika App and touch on the desire for intimacy via AI psychology, revealing a crucial aspect of power central to this discourse. Even if one only produces an entity that can fool a substantial plurality of human beings into thinking it is conscious, that itself is an impressive milestone in the instantiation and psychology of power. One of the questions that distributed consciousness discourse is forcing human beings to confront is: What does this impulse to create awareness say about the people engaged in the process? If we may not ever truly know why we as a species are so enduringly fascinated with inorganic sentience, we may at least be able to fathom the psyches of those who actively pursue this dream and disentangle their healthy and unhealthy personal rationales for undertaking the quest. Thus, cybernetics can become a window on the question of ethics.

Our own work has attempted to come to terms with the reasons for seeking to create other minds. For example, *Seek serenity and meaning, leave the work to us* (2016–18) is a work that exists as part of the series *Spiritual Reality*. The premise of these works is to consider a potential world where human labor has become superfluous—even a kind of obstacle—to new models of production. The tension between the advance of automation and the diminution of the sense of self that many people derive from their work is already present in contemporary narratives about the "disappearance" of jobs and the "replacement" of workers by algorithms. That such dynamics are accelerating is all but unarguable, and the *Spiritual Reality* project seeks to explore the cultural responses that will emerge in order to manage this advancing economic and psychological crisis. *Oracle (Spiritual Reality)* (2020) is a chatbot who explains the ways in which human beings have been integrated into a culture where they are not only no longer the central engine of economic production but also no longer the presumed primary subject of political organization. The chatbot in the work recounts the painful process of replacement but offers a hopeful message: in a world where humans don't have to do anything, at least they can do anything they want.

I am Made and Remade Continually, *The Waves of Possibility from The Ocean of Reality*, and *Rolling Over The Wave Will Shoulder Me* (all 2018) also integrate questions about the limits and potentialities of expanded consciousness and permeable bodies, influenced by Astrida Neimanis's essay "Hydrofeminism: Or, On Becoming a Body of Water" (2012).[15] Neimanis' ideas, like the BwO, emphasize flow, but also leakage, permeability, and the role that context plays in shaping our perspectives. The works, consisting of inflatable mattresses for swimming pools emblazoned with text, playfully integrate the notions of fragility and flow present in Neimanis's concept. Indeed, water, when looked at from a hydrofeminist perspective, could well be said to be as "alive" as the creatures who cannot survive without it. Perhaps the idea of water as a form of consciousness is a radical step and would require seeing consciousness as one of the primitives (e.g., mass, charge, spin, etc.)—though so many ancient traditions have already integrated notions of sentience into aspects of the natural world, from the sea and rivers to the sun and mountains. Whatever one may feel about the ultimate consciousness of water, the idea of an expanded understanding of consciousness is increasingly becoming inescapable. Indeed, one may wonder why it has taken us so long to come back to this notion that so many indigenous traditions have accommodated.

It is somewhat ironic that discourses around AI have led us back into a greater awareness of the possibility of consciousness in other beings and to the instabilities of our own sense of our physical and psychic integrity. Our longstanding interest in intersubjectivity requires a commitment to inquiry, even if one is not always certain where that inquiry will lead. We may learn more by being uncertain than by being overly sure of ourselves. If there is, as the Leonard Cohen song runs, a crack in everything and that is "where the light gets in," perhaps it is the goal of art to open more fissures in the world to allow illumination to come from unanticipated sources. David Chalmers is quoted in Thornhill's aforementioned *Financial Times* article emphasizing

15 Astrida Neimanis, "Hydrofeminism: Or, On Becoming a Body of Water," in *Undutiful Daughters: New Directions in Feminist Thought and Practice*, ed. Henriette Gunkel, Chrysanthi Nigianni, and Fanny Söderbäck (New York: Palgrave Macmillan, 2012), 85–100.

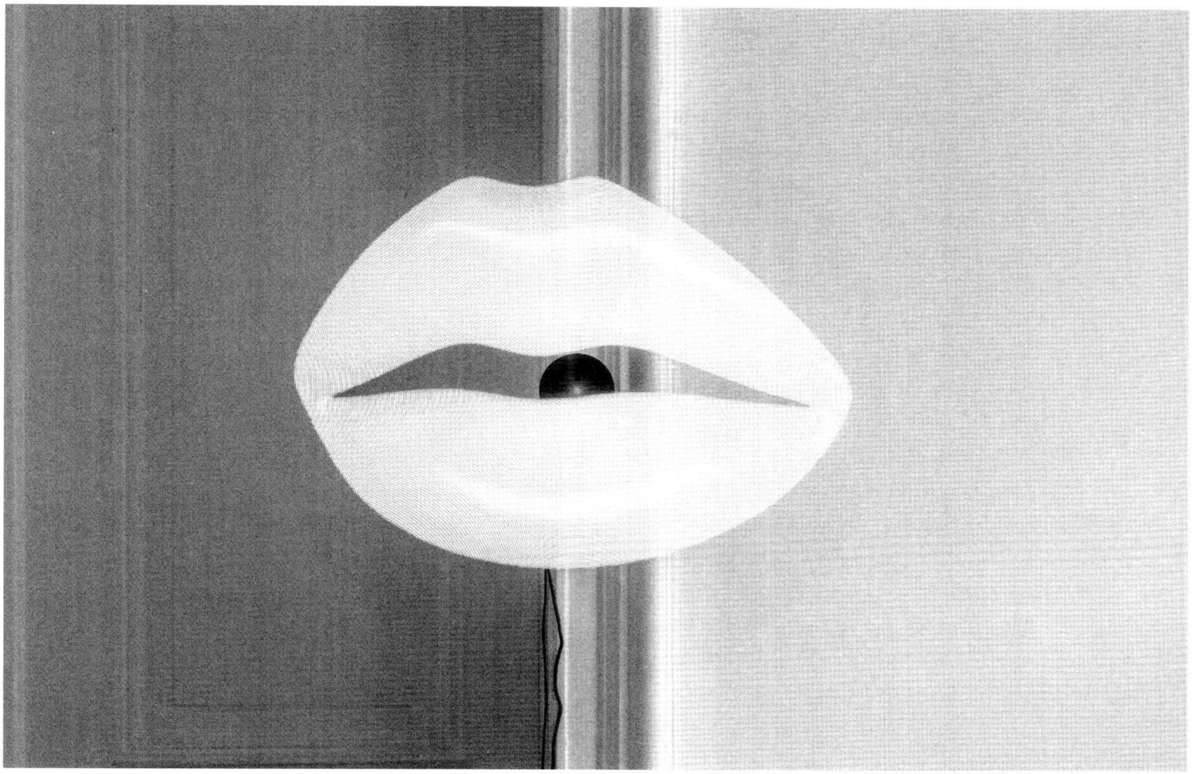

Lou Cantor, *Oracle*, 2020. Holographic fan. EMAF "First Person Plural," 2020

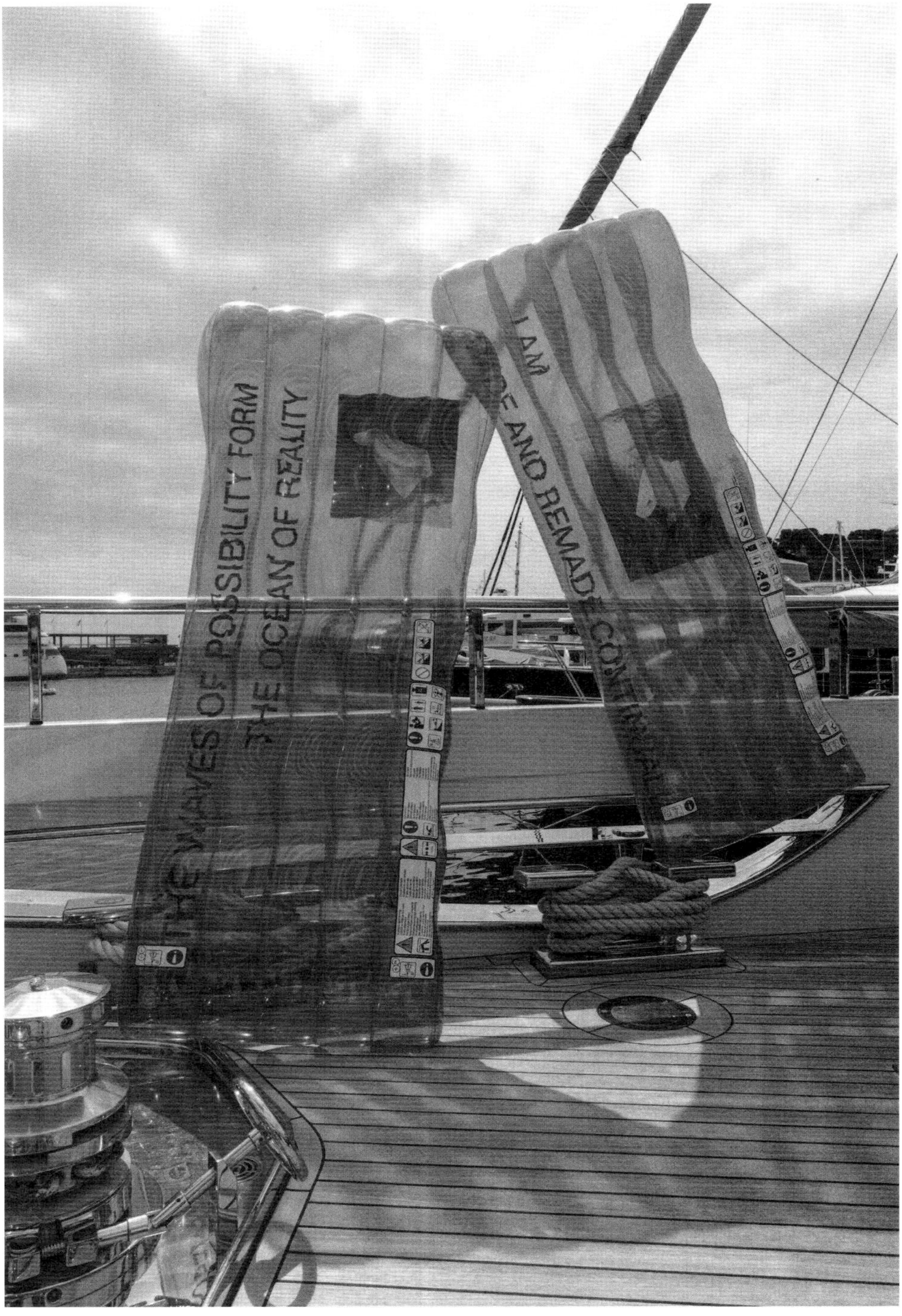

Lou Cantor, *The Waves of Possibility from The Ocean of Reality* and *I am Made and Remade Continually*, 2018. Air mattress, digital print on TPU foil. “Hydrofeminism,” 2018

the importance of erring on the side of sentience when interacting with entities like GPT-3. This, Chalmers says, should have equal implications for our relationships with other living beings with whom we (often begrudgingly) share the earth: "I am open to the idea that a worm with 302 neurons is conscious, so I am open to the idea that GPT-3 with 175 billion parameters is conscious too."[16] Though the evolving ecology of consciousness remains highly uncertain, the impulse Chalmers expresses toward openness and the embrace of new possible forms of awareness existing in the world is a salutary one for a moment when many are all too eager to draw exact and exacting borders over the "blooming, buzzing confusion" of human experience, to use the words of the philosopher William James.[17] Looked at from one perspective, drawing a line could be seen as the creation of a limit; looked at from another, it could become a means of connection. As we seek to engage other minds, other forms of awareness, we must bear in mind the quantum dimension of the lines we draw, and invest our efforts in making visible the plenitude of the latter interpretation.

16 Thornhill, "Is AI Finally Closing In on Human Intelligence?"

17 William James, *Principles of Psychology* (New York: Holt, 1890), 1:488.

Written for a male actor
Direct, neutral voice, a little clipped/robotic/helpful/upbeat
(All of the "Goodbyes" should be a little reminiscent of the AOL logoff "Goodbye")

Music for a few seconds before speech comes in.

1_BadDadMenu: *(knowing advice)* Fathers have become an important part of everyday life, so when the unit we depend on malfunctions it's easy to panic. The truth is that many problems also have an easy fix...Why don't you begin by telling me the nature of your father's problem. Press 0 for software, 1 for hardware.

2_SoftwareMenu: Has his problem reporting or capacity to be self-critical stopped working? Press 0 for yes, 1 for no.

3_YesSelfCriticalStoppedWorkingMenu: Does he have a tendency to overheat and become violent, acting on terrible thoughts such as, "One should let one's nails grow for a fortnight. O, how sweet it is to drag brutally from his bed a child with no hair on his upper lip and with wide open eyes, make as if to touch his forehead gently with one's hand and run one's fingers through his beautiful hair. Then suddenly, when he is least expecting it, to dig one's long nails into his soft breast, making sure, though, that one does not kill him; for if he died, one would not later be able to contemplate his agonies. Then one drinks his blood as one licks his wounds; and during this time, which ought to last for eternity, the child weeps"? Press 0 for yes, 1 for no.

4_YesTendencyToOverheatMenu: Has he eaten you or any of your siblings? Has he deleted files without your permission? Press 0 for yes, 1 for no.

5_YesEatenSiblingsMenu: A guilty conscience is the number one liar and the producer of suspicion. If your dad presents himself as a victim in order to avoid guilt, press 0. If his spyware is outdated, press 1.

6_PresentsHimselfAsVictimMenu: Reconnect your father to the appropriate networks. If you are unable to connect him, press 0. If he perverts the network, press 1.

7_UnableToConnectEnd: Attempting a physical memory dump...*(grunt)*...Incorrect password. Your password must be at least 18770 characters and cannot repeat any of your previous 30689 passwords. Goodbye. *(All of the "Goodbyes" should be a little reminiscent of the AOL logoff "Goodbye")*

8_PervertsNetworkEnd: *(angry modem sound) (quickly and without affect)* bugs, chugs, drugs, hugs, jugs, lugs, lugs, lugs, lugs, mugs, plugs rugs shrugs slugs, thugs, tugs, tugs, tugs, tugs, tugs, tugs, tugs, tugs, tugs *(angry modem sound)*

9_SpywareOutdatedMenu: If he overexpresses himself or his text is too large, press 0. If his files open with the wrong program and he can't communicate without complex prompts, press 1.

10_OverExpressesEnd: The earth has many keys. Where melody is not is the unknown peninsula. All things swept sole away. This—is immensity—...Goodbye.

11_FilesOpenWithWrongProgramEnd: Processes of generation; deeds of settlement. The urge to marry well; wit to invest in the properties of healing-springs. Our children and our children's children, o my masters...Goodbye.

12_NoEatenSiblingsMenu: Does your father let strangers use your identity and turn all they touch to gold? Press 0 for yes, 1 for no.

13_YesLetStrangersUseYourIdentityMenu: If he has encountered a catastrophic failure, press 0. Press 1 for other bugs.

14_CatastrophicFailureEnd: Uh oh! Abort, retry, or ignore. Goodbye.

15_NoLetStrangersUseYourIdentityMenu: Has he been infected by a porn virus? Press 0 for yes, 1 for no.

16_YesInfectedbyPornVirusEnd: Banish your father to Tartarus. Goodbye.
***(*Tartarus is pronounced tahr-ter-us*)

17_NoInfectedbyPornVirusEnd: Runaway processes take up all of the processors' cycles. The usual suspects are badly written device drivers and legacy software installed on a newer operating system. Sometimes a hard reboot is required. Goodbye.

18_NoTendencyToOverheatMenu: Does he exhibit control freakiness? Press 0 for yes, 1 for no.

19_YesControlFreakinessMenu: Is he drunk all the time? Press 0 for yes, 1 for no.

20_YesDrunkAllTheTimeMenu: Check his menu. Press 0 if *(repetitive computer glitch sound)* 1 for *(repetitive computer glitch sound)*.

21_ZeroGlitchEnd: Please contact a technical support group for further assistance. Goodbye.

22_OneGlitchEnd: Try to *(repetitive computer glitch sound)*. Goodbye.

23_NoDrunkAllTheTimeMenu: Have you tried transferring his user account? Press 0 for yes, 1 for no.

24_YesTriedTransferringHisUserAccountEnd: Back up all data and prepare for him to fail soon. Goodbye.

25_NoTriedTransferringHisUserAccountEnd: Please contact your system administrator or a technical support group for further assistance. Goodbye.

26_NoControlFreakinessMenu: Is he unable to open help and support? Press 0 for yes, 1 for no.

27_YesUnableToOpenHelpAndSupportMenu: Does he have a mistress or has he been subject to sorcery, divine malice, trojan horses, scareware, worms, or other malware?

28_NoMistressEnd: Quit and restart while pressing command T. Login to the Linksys and upload The Extramarital Narcissist, The World of the Narcissist, and The Pathology of Love. Goodbye.

29_NoUnableToOpenHelpAndSupportMenu: If you have update software available, press 0, otherwise press 1.

30_UpdateSoftwareAvailableEnd: Browse to HKEY underscore CURRENT underscore USER backslash Software backslash Classes backslash Local Settings backslash Software backslash CurrentVersion backslash TrayNotify, and delete both IconStreams and PastIconsStream values. Goodbye.

31_NoSelfCriticalStoppedWorkingMenu: What type of code does he use to communicate with you? Press 0 for Limbo, Lasso, Mercury, Mouse, or Yorick. Press 1 for Executable UML, Draco, C++, Janus, Python, or Zeno. ***(C++ is pronounced "See plus plus")*

32_LimboLassoMercuryNoSelfCriticalMenu: Is he a somatic narcissist who compulsively hunts for new feminine prey to sacrifice on the insatiable altar of his False Self or otherwise run into fatal errors when executing simple scripts?

33_YesSomaticNarcissistMenu: Does he pay more to avoid taxes than the sum he would be taxed and automatically reject cookies? Press 0 for yes, 1 for no.

34_YesPayMoreToAvoidTaxesMenu: Is he a Tartuffe or otherwise subject to peer-to-peer sharing issues? Press 0 for yes, 1 for no.

35_YesTartuffeEnd: Perhaps he requires better ventilation. Check if there is dust in the fan that cools the processor, then slip him a Xanax. Goodbye.

36_NoTartuffeEnd: We are sorry to hear that you are disappointed with your connection but we can assure you that your subscribed connection is efficient. Goodbye.

37_NoPayMoreToAvoidTaxesMenu: Does he sacrifice children to Moloch and perform unexpected reboots? Press 0 for yes, 1 for no. *(Moloch is pronounced Moloc)*

38_YesSacrificeChildrenEnd: Disable password-protected filesharing and check that your firewall has placed your network in a trusted zone. Goodbye.

39_NoSacrificeChildrenEnd: Please read Ethical Relativism and Absolute Taboos to better understand the nature of the problem. Goodbye.

40_NoSomaticNarcissistMenu: Has he fathered more than 9 children by more than 4 women? Has his help and support stopped responding? Press 0 for yes, 1 for no.

41_YesFatheredMoreMenu: Has his rootkit allowed malicious software to violate you? Press 0 for yes, 1 for no.

42_YesRootkitViolateEnd: Try going to Safe Mode. Goodbye.

43_NoRootkitViolateEnd: Check your motherboard for short circuits. Goodbye.

44_NoFatheredMoreMenu: Does he get stuck in infinite or unproductive loops that shut you out? Press 0 for yes, 1 for no.

45_YesStuckInfiniteLoopsEnd: With MenuOrder selected, choose Edit, then Permissions. Click the Advanced button and remove the tick next to "Inherit from parent the permission entries..." When prompted click Copy and select OK, to close the Advanced Security Settings window. If this doesn't work, explicitly threaten to abandon him. Goodbye.

46_NoStuckInfiniteLoopsEnd: One way to increase performance is by turning off some of the financial effects that make your father seem cool but use valuable system resources. Goodbye.

47_ExecutableUMLDracoNoSelfCriticalMenu: Does he wear a mask of kindness that takes, not gives, that wants, not offers, that awakes, not pacifies? Has a system call that should never fail truly failed? Press 0 for yes, 1 for no.

48_YesMaskOfKindnessMenu: Do constant pop-ups prevent you from having a real discussion with him? Press 0 for yes, 1 for no.

49_YesConstantPopUpsMenu: Has he killed all of his crew members and abandoned you to join the Sith? Press 0 for yes, 1 for no.

50_YesKilledAllCrewEnd: The vegetable academy of silences...your name that sounded like poppies...the ponds...my going home...the crazy priest who went out of his mind during Mass...Waters...Goodbye.

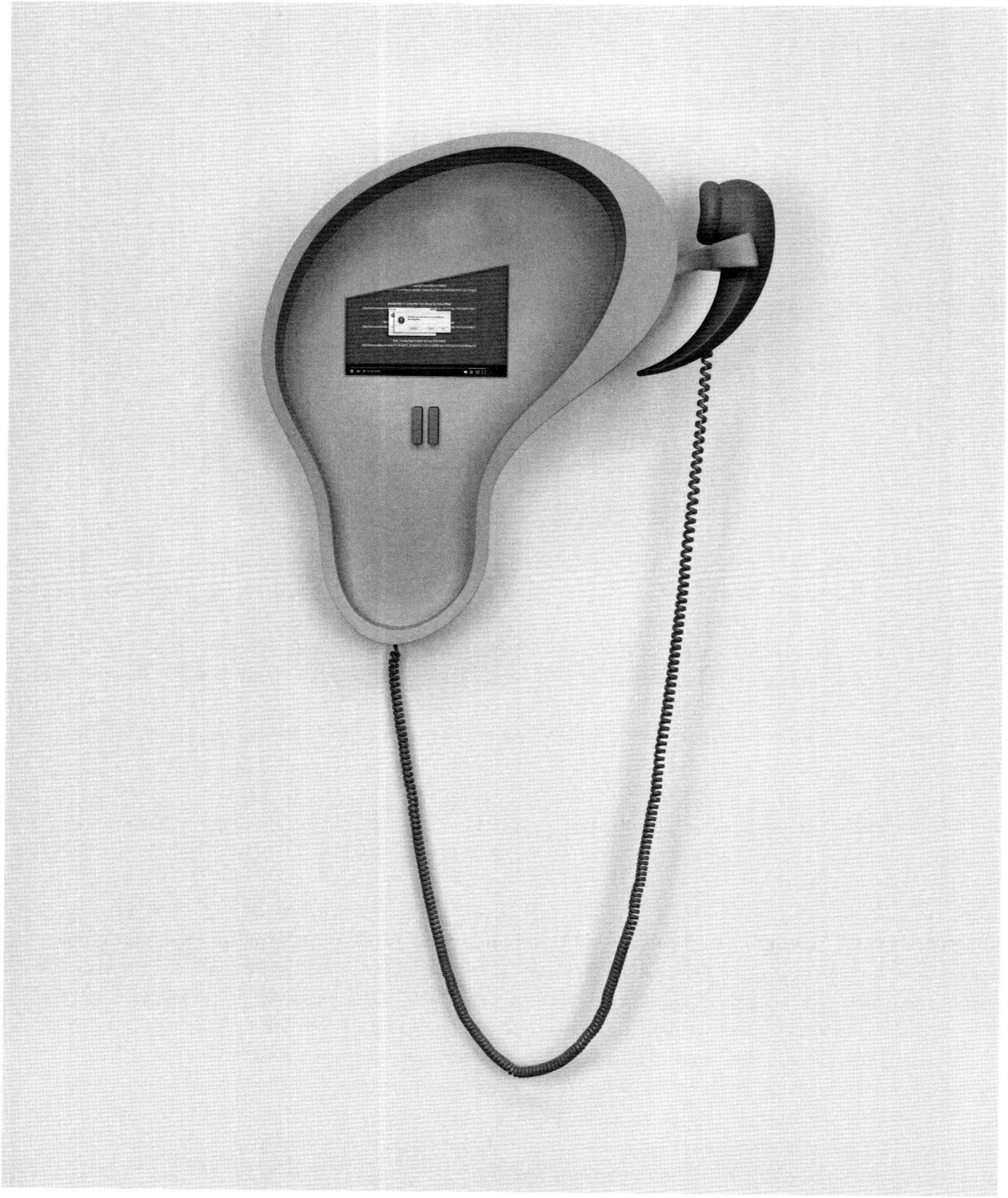

Camille Henrot, *Bad Dad & Beyond*, 2015, from the series *Interphones.* Interactive sculpture, mixed media

51_NoKilledAllCrewEnd: Type DisableRemovable into the New Value number one box, then press Enter and double-click DisableRemovable. Change its value to one and restart. Goodbye.

52_NoConstantPopUpsMenu: Has he stopped responding to text input and disowned you for declining to indulge his vanity? Press 0 for yes, 1 for no.

53_YesStoppedRespondingToTextEnd: It takes two to tango. Stop playing along with his schemes and repair drive partitions. Goodbye.

54_NoStoppedRespondingToTextEnd: If the system allows, adding more RAM can often increase introspection. Goodbye.

55_NoMaskOfKindnessMenu: Is his application manifest running improperly? Has he tried to sacrifice you as a burnt offering? Press 0 for yes, 1 for no.

56_YesApplicationManifestRunningImproperlyMenu: Has he compensated for his own resource leaks by forcing you into a marriage that will net him property? Press 0 for yes, 1 for no.

57_NoCompensatedForResourceLeaksEnd: Go and uncover the great Sphinx. Goodbye.

58_NoApplicationManifestRunningImproperlyMenu: Has he given in to madness and despair, resulting in deadlock? Press 0 for yes, 1 for no.

59_YesGivenInToMadnessEnd: Refer to motherboard manual for error code meaning. Goodbye.

60_NoGivenInToMadnessEnd: Isolate drive on primary and gouge your eyes out. Goodbye.

61_HardwareMenu: Is he obsessed with failure? Does he regard other men as nothing but puppets of a sort, created to fill up an empty world, dividing them into two classes: those he greets because some chance has put him in contact with them, and those he does not greet? Are both these categories of individuals equally insignificant in his eyes regardless? Press 0 for yes, 1 for no.

62_YesObsessedWithFailureMenu: Has he been affected by the Corrupted Blood bug or attempted to murder you or your siblings in order to alter weather patterns? Press 0 for yes, 1 for no.

63_LimboLassoMercuryCorruptedBloodMenu: Does he open and close programs at random, treating you as an object, an extension of himself, devoid of a separate existence and denuded of distinct needs? Press 0 for yes, 1 for no.

64_YesOpenAndCloseProgramsMenu: Is he so obsessed with his power supply that he sold you or any of your siblings? Press 0 for yes, 1 for no.

65_NoObsessedWithHisPowerSupplyEnd: Be sure to back up your files on a regular basis, but refuse him entry to your premises. Goodbye.

66_NoOpenAndCloseProgramsMenu: Does he display error messages to deny paternity? Press 0 for yes, 1 for no.

67_NoDisplayErrorMessageEnd: Accept that we don't choose our parents and do your best to disable BIOS memory options such as caching or shadowing. Goodbye.

68_ExecutableUMLDracoCorruptedBloodMenu: Doth the sin of self-love possess all his eye and all his soul and all his every moving part, so grounded inward in his code he thinks no display so gracious as his own, no shape so true, no truth of such account? Press 0 for yes, 1 for no.

69_YesSinOfSelfLoveMenu: Is his clock out of sync? Press 0 for yes, 1 for no.

70_NoSinOfSelfLoveMenu: Does his system freeze if you don't constantly update him? Press 0 for yes, 1 for no.

71_NoSystemFreezeEnd: It is imperative that you attend to your well-being by spending less time in front of your father, getting outside, eating healthily, sleeping, staying free from destructive behaviors, exercising, and being able to love yourself even if your father is unable to. Goodbye.

72_NoAffectedByCorruptedBloodMenu: Does he make strange noises and call down curses on you? Does he want to give your older brother dominion over you? Press 0 for yes, 1 for no.

73_YesStrangeNoisesMenu: Are his logic gate inputs faulty, resulting in racism? Press 0 for yes, 1 for no.

74_YesLogicGateInputsFaultyMenu: Is he stuck in Caps Lock? Press 0 for yes, 1 for no.

75_NoCapsLockEnd: Do not respond to his pleading, nostalgic, flattering, or threatening error messages. Goodbye.

76_NoLogicGateInputsFaultyMenu: Does he open files at random, as though he's possessed? Press 0 for yes, 1 for no.

77_YesOpenFilesAtRandomEnd: Choose factory data reset from the Backup and Reset settings, then reboot. A negative relationship with your father will only come to define you if you allow it to. Goodbye.

78_NoOpenFilesAtRandomEnd: Swap power lead. Isolate on controller. Power management. Goodbye.

79_NoStrangeNoisesMenu: Has he trapped you in a labyrinth after attempting to divide by zero? Press 0 for yes, 1 for no.

80_YesTrappedYouInLabyrinthMenu: Is he in the middle of his life-cycle and adding useless upgrades? Press 0 for yes, 1 for no.

81_NoTrappedYouInLabyrinthMenu: Are his Basic Input Output System failures due to alcoholism? Press 0 for yes, 1 for no.

82_NoBasicInputOutputSystemEnd: Isolate drive on primary and be careful not to visit him on special occasions, or in emergencies. Goodbye.

83_NoObsessedWithFailureMenu: Does he fear loss of control and freeze up when you have several programs running? Press 0 for yes, 1 for no.

84_YesFearLossOfControlMenu: Has he stopped you from connecting with friends and imprisoned you within your mother or your home? Press 0 for yes, 1 for no.

85_YesStoppedYouFromConnectingMenu: Is he controlled remotely by Cow-Eyed Hera, She of the Heights, Goat-Eater of the White Arms, of the Mound? Press 0 for yes, 1 for no.

86_YesControlledRemotelyMenu: Does he corrupt all files created after a certain date? Press 0 for yes, 1 for no.

87_NoControlledRemotelyMenu: Does he expect preferential treatment from others, expect others to cater to his needs, without being considerate in return? Press 0 for yes, 1 for no.

88_NoStoppedYouFromConnectingMenu: Is he obsolete? Press 0 for yes, 1 for no.

89_YesObsoleteMenu: Can you uninstall? Press 0 for yes, 1 for no.

90_YesUninstallEnd: Try running a defrag. Goodbye.

91_NoUninstallEnd: Don't make excuses for him. Don't try to understand him. Do not empathize with him—quit him immediately. Goodbye.

92_NoObsoleteMenu: Is he lazy? Press 0 for yes, 1 for no.

93_YesLazyEnd: A failure is impending. Goodbye.

94_NoLazyEnd: We advise you to be patient because patience is the best cure for the rainbow wheel of death and the blue screen of death. It has benefits in this world and in the Hereafter. Goodbye.

95_LimboLassoMercuryNoFearLossOfControlMenu: Is he cold and calculating? Press 0 for yes, 1 for no.

96_YesPromisesMenu: Is his sorting algorithm malfunctioning, leading him to try to top your feelings with his own stories? Press 0 for yes, 1 for no.

97_YesSortingAlgorithmMalfunctioningEnd: Try restarting him. If this doesn't work, explicitly threaten to abandon him. Goodbye.

98_NoPromisesMenu: Does his dataflow go squiggly when you talk about your feelings? Press 0 for yes, 1 for no.

99_NoDataflowSquigglyPromisesEnd: Vice is the passion of failure. Goodbye.

100_ExecutableUMLDracoNoFearLossOfControlMenu: Does he make promises and then delete them from his memory? Press 0 for yes, 1 for no.

101_YesColdAndCalculatingMenu: Is his system fault-tolerant? Press 0 for yes, 1 for no.

102_YesFaultTolerantEnd: A geometry problem may be placing undue stress on your motherboard. Don't make excuses for it. Don't try to understand it. Do not empathize with him—quit him immediately. Goodbye.

103_NoFaultTolerantEnd: Your father may require professional debugging. Goodbye.

104_YesDataflowSquigglyColdAndCalculatingEnd: Mirror his actions, then debug when he gets upset. Goodbye.

After the deep learning revolution, which came to fruition in the early 2000s (but started way earlier with the design of deep, feedforward, multilayer artificial neurons called "perceptrons" in 1967), the appetite for a living machine increased. Today, many top scientists linking biology with computation, like Karl Friston[1] or Manolis Kellis,[2] are discussing the intersection of artificial intelligence and synthetic biology as the next frontier. Many are also asking: Why is AI progressing so slowly? Despite remarkable work performed by various neural networks in natural language processing and image recognition, we are still hesitant to call these networks intelligent, even in a narrow sense. What is missing?

The obstacles that computational programmers are encountering today demonstrate that computational theory still isn't sufficient for constructing a conscient machine in the human image. The inner workings of a living, thriving, psychophysical, and conscious unit can't be thoroughly explained using the wide-ranging, abstract laws of data processing.

In the 1980s, the general understanding of the AI paradigm was subverted. The then prominent metaphor of the mind as a computational machine (based on the concept of the Turing machine) was meant to explain the processes of human cognition, but it did so within a rather narrow scope. The metaphor was most useful in describing computational operations, inductive and deductive methods of obtaining answers, and simple calculations (e.g. solving algorithmic problems) in situations with unequivocally certain solutions. But other cognitive processes are more complex, such as intuitive approaches to proving the accuracy of hypotheses, drawing conclusions with unseen or unacknowledged presumptions, creative (non-algorithmic) discovery, and, evidently, the plethora of living actions and interactions in the world. Faced with these hurdles, the computational paradigm was downgraded by brothers Stuart and Hubert Dreyfus to a theory that explored the problem of how the mind, body, and consciousness relate to one another and to human experience, by way of material proof that was embodied in the imaginary of a truly intelligent machine.[3]

The paradigm was downgraded, but it wasn't disregarded. It continued to evolve and undoubtedly found its application in a number of disciplines like optimization, operational research and planning, as the community of growing AI experts determined its efficacy in those areas to be of value. Yet, in the scientific pursuit of answers regarding consciousness and mental processes, computational concepts did not provide satisfying answers. Thus, a concurrent paradigm began to gain speed. Connectionist theory, which came out of the neuropsychology theories of Donald Hebb, provided a subsymbolic paradigm of artificial intelligence that assumed that the passage from dispersed, decentralized, and embodied knowledge to individual

1 David Holmes, "Karl Friston: Resisting the Philosopause," *Lancet Psychiatry* 1, no. 2 (July 2014): 117, https://doi.org/10.1016/s2215-0366(14)70292-5.

2 Susan E. Celniker et al., "Unlocking the Secrets of the Genome," *Nature* 459, no. 7249 (June 2009): 927–30.

3 Hubert Dreyfus and Stuart Dreyfus, "Why Computers May Never Think Like People," in *Knowledge Management Tools*, ed. Rudy L. Ruggles III (Newton, MA: Butterworth-Heinemann, 1997), 31–50.

4 Jane Cull, *Living Systems: An Introductory Guide to the Theories of Humberto Maturana & Francisco Varela* (self-pub., CreateSpace, 2017).

subjective knowledge was illusory. The goal of connectionist theory was to revise Cartesian theory about how the outside world was reflected in cognitive actions. According to the connectionists, the reflection of the world through formal systems is the only method of generating intelligent behavior. Connectionist work in general does not need to be biologically realistic, but the very structure of artificial neural networks is derived from that of biological ones, and this parallel in low-level structure is often argued to be an advantage of connectionism in modeling cognitive structures compared with other approaches.

Rather than build computer programs using traditional programming languages, which are based on logical inferences and conditional instructions, connectionists create a simulation of a given process through artificial neuron networks. Such a network is made up of a number of individual units that function similarly to human nerve cells. The network assumes that the brain processes information using a large number of small-scale functional units that build interconnected webs that are similar to functional neural networks made up of nerve tissue. The connectionist architecture doesn't possess a single, central processor to carry out operations but, rather, is made up of a series of neurons transferring impulses among one another and in tandem within an entire network of connections. The strength of each connection is described by a weight, which is altered in the process of learning the network.

Connectionists place an emphasis on the importance of modeling according to the same rules that govern human cognitive processes. For them, this means intelligence comes out of the interaction between numerous basic units that process information. In this approach, a computer isn't treated as itself a potential mind but as a research tool: a way to understand the human mind by attempting to recreate it. Crucially, a neural network isn't programmed to complete a specific task, but, nonetheless, it achieves its goals through a method of trial and error, "learning" by obtaining the information available upon entry. In solving a cognitive challenge, it makes mistakes, much like a human does.

Though the connectionist approach is complex and nuanced, it is still based on the fundamental idea that human consciousness can be recreated mechanically. This notion has been contested by cognitive scientists including Francisco Varela and Humberto Maturana.[4] Critics of the computational paradigm, particularly those oriented toward phenomenology such as Hubert Dreyfus, argued that a "computerized" mind can't be compared to the human processes of learning and creating. Crucially, AI is not capable of independently formulating meaning in a non-algorithmic fashion, a method that favors a view of data processing based on a program that is available at hand. The essential, classic, computational theory of consciousness doesn't assume an AI is capable of creative learning. This classic approach thus only works for simple operations of deductive or syllogistic reasoning, but doesn't touch upon learning understood as adapting to the changing environment at all.

Human beings feature a system far more complex than either model offers, one that encompasses sight, perception, and thought. Dreyfus pointed out that sensory experiences and abstract thinking both take place via the mind as well as the body.

Neither the computational nor the connectionist models take into account the relationship between the mind and the body, failing in this regard to answer one of the key aspects of consciousness and the human condition overall. Most of the criticism of the computer metaphor is lodged at its project of rationality and related "technicalization" of humankind. But the computational paradigm has several other issues. It doesn't factor in the complexity of the human system, which carries out so many life functions of a strictly cognitive-practical character, such as perception. As a result, in both the realms of philosophy and artificial intelligence, a number of alternative, phenomenological ideas began to germinate, rooted in the belief that in the mind, and certainly in the body, there are much more complex and diverse mechanisms of observation, processing, and exploiting information at work.

These ideas arose from the foundations of cybernetics, as a systemic approach to human cognitive abilities primarily made possible thanks to the contributions of social scientist Gregory Bateson.[5] Bateson, a pioneer of systems theory, clearly states that the role of information in human actions, as in those of every complex organism, is the particular signal coming from the environment, which becomes the stimulus for a particular behavior (either practical or cognitive). This difference in action is a result of another difference between the given stimulus and other possible stimuli that had been previously received and recorded into its standard repertoire of activities. When talking about the role of information in human actions, we can say that it appears only after it is linked to this difference and shift in signals. Registered at the anatomical and functional levels, these differences then serve as models and schema for interpreting new and unfamiliar phenomena that have not been previously encountered; this is how perception functions, as well as memory, speech, and abstract thinking.

This means that information is derived from the environment and is not an independent entity that a human being comes in contact with. For humans, information is contained within external signals, which are processed via the schemas already present in the mind and body. Such processes alter the structural makeup of the nervous system, which extends far beyond the brain.

Recent research in the field of neurophysiology has changed the way we see the brain: it's

5 Gregory Bateson, *Mind and Nature: A Necessary Unity* (New York: Dutton, 1979).

an automatized system that functions through reflexes and conditioning, moving toward a closed system producing its own understanding of the outside world. In such a system, the introduction of a sensory stimulus is no longer synonymous with information but is rather a catalyst for changes within the brain.

The conclusion in all this is that the functioning of information within the human mind is characterized by a multiplicity of layers and their many aspects. Nothing has yet indicated that this complex functionality might be described, as it is in computational theory, with the simple phrase "data processing." Information, in the cognitive and practical aspects of human activity, doesn't simply boil down to signals gleaned from the environment or stimuli acting upon the body. It isn't the sum of these signals recorded in matter (in the flesh, speech, writing, print, image). It is only after a whole grouping of individual meanings come together to build up the rich store of meaning that information manifests itself within the scope of human existence.

The anti-computational model of cognition, proposed, among others, by Francisco Varela, Alva Noe, Ezequiel di Paolo, and Hanne de Jeagher, presupposes humankind as an informational-cognitive system that isn't closed or unalterable but open and autonomous. Every theory that considers the physiology of cognitive processes within the body presents the human as a so-called active explorer. It's a particular system of perception that actively relates to all stimuli, making their way to the organism from its environment, which then registers changes and processes these changes into information. Further, computationalism takes the computer as its dominant metaphor, while in physiological cognitive processes, that metaphor is replaced by a view of structural relations and connections within the environment, as well as of activities that stimulate the cognitive process. In computationalism, cognition is given a passive character, which can be explained through a study of internalized processes. However, in the psychological paradigm used to explain cognition, the process is an active one, and its careful study is meant to determine the mutual correlations between the body, mind and the environment.

In the coming years we will probably witness several attempts to go beyond narrow AI. Connectionism, currently developing rapidly through major advancements in various types of neural networks, could be combined with some form of yet unknown embodied cognition. It remains to be seen whether they will result in any form of technological physiology that will allow for at least minimal affective states. Perhaps work on projects such as xenobots—synthetic organisms designed by computers to perform some desired function and built by combining together different biological tissues—marks the very beginning of some form of artificial life itself.

Lynn Hershman Leeson is a pioneering media artist and filmmaker who has been working with emerging technologies since the 1970s. She's been based in San Francisco for her entire career, and her proximity to Silicon Valley and several research institutions has allowed her to develop a keen awareness for changes in scientific research. As research priorities have shifted from a focus on digital technologies and the World Wide Web to advancements in biotechnology and genetics, Hershman Leeson's attention has also moved. However, she has remained dedicated to collaborating with leading computer scientists, technologists, biologists, geneticists, artificial intelligence developers, and those working at the forefront of surveillance technology. At the heart of her work lies an interrogation of human subjectivity in the age of machines, and an exploration of how identity is formed, as well as intimacy mediated, through technology. Below Hershman Leeson talks with Samantha Ozer about how her recent work with DNA *relates to her early video works, and about the connection between one's digital and biological selves, which has been pulled into focus during the coronavirus pandemic.*

SAMANTHA OZER: One of your earliest and longest-running works, *The Electronic Diaries* (1984–2019), is a series of vulnerable, intimate, confessional videos in which you speak directly into a camera to create a video diary. This process of archiving the self, using a home video camera that we may now consider a fairly archaic device, concluded with the diaries themselves being archived in DNA. While DNA is a genetic material, its four-letter nucleotide code operates akin to the digital binary code used by computers and other electronic. Scientists at the Wyss Institute for Biologically Inspired Engineering at Harvard University have also pointed out that DNA storage is more durable than existing silicon-based storage media. While DNA digital storage is a more stable method for archival material and might be the future of digital storage, currently, it is too costly. It also brings up ethical questions concerning the effects that advancements in biotechnology are having on intersubjective relations, and on the bounds of the self. I view your recent work with DNA, *Infinity Engine* (2014), as being diaristic, if not poetic.[1] Considering that you've spoken about your work with DNA as being a haiku of your life, I'm curious if you also see this work as diaristic?

LYNN HERSHMAN LEESON: I haven't thought about DNA as being intimate; however, it really is the most intimate thing inside of ourselves. I didn't have a conceptual strategy when creating *The Electronic Diaries*; rather, I wanted to teach myself how to use the camera and edit, so the diaries became a strategy for learning. The project also took the form of a diary as I didn't have anybody else to shoot, so I had to speak to the camera myself. One thing that surprised me was that the more I got into my own truth, the more people

1 The *Infinity Engine* is a multimedia installation that is a functional replica of a genetics lab. The installation is composed of a series of rooms, with each highlighting a specific ethical or technical aspect of genetics research. The final room, "Room 8: DNA," is a locked lab door, with a small window for public viewing, that contains one vial with the powder of the LYNN HERSHMAN antibody and another with three hundred nanograms of DNA containing the digital documentation of the entire exhibition, as well as *The Electronic Diaries* (1986–94). The LYNN HERSHMAN antibody was created by putting together the amino acids that numerically correspond to Hershman Leeson's name to form a unique antibody chain.

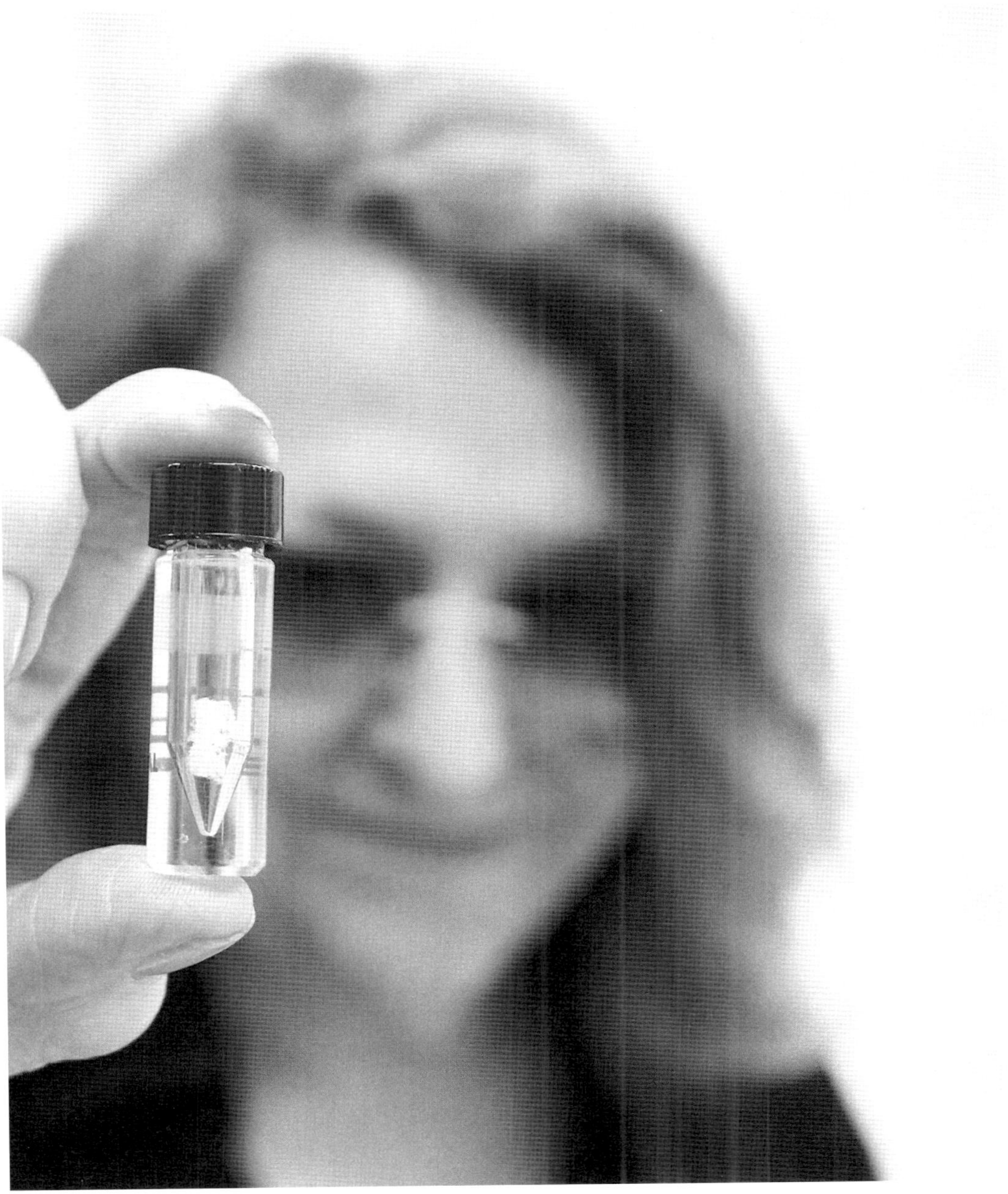

Lynn Hershman Leeson holding the *LYNN HERSHMAN Antibody*, 2017. Archival digital print

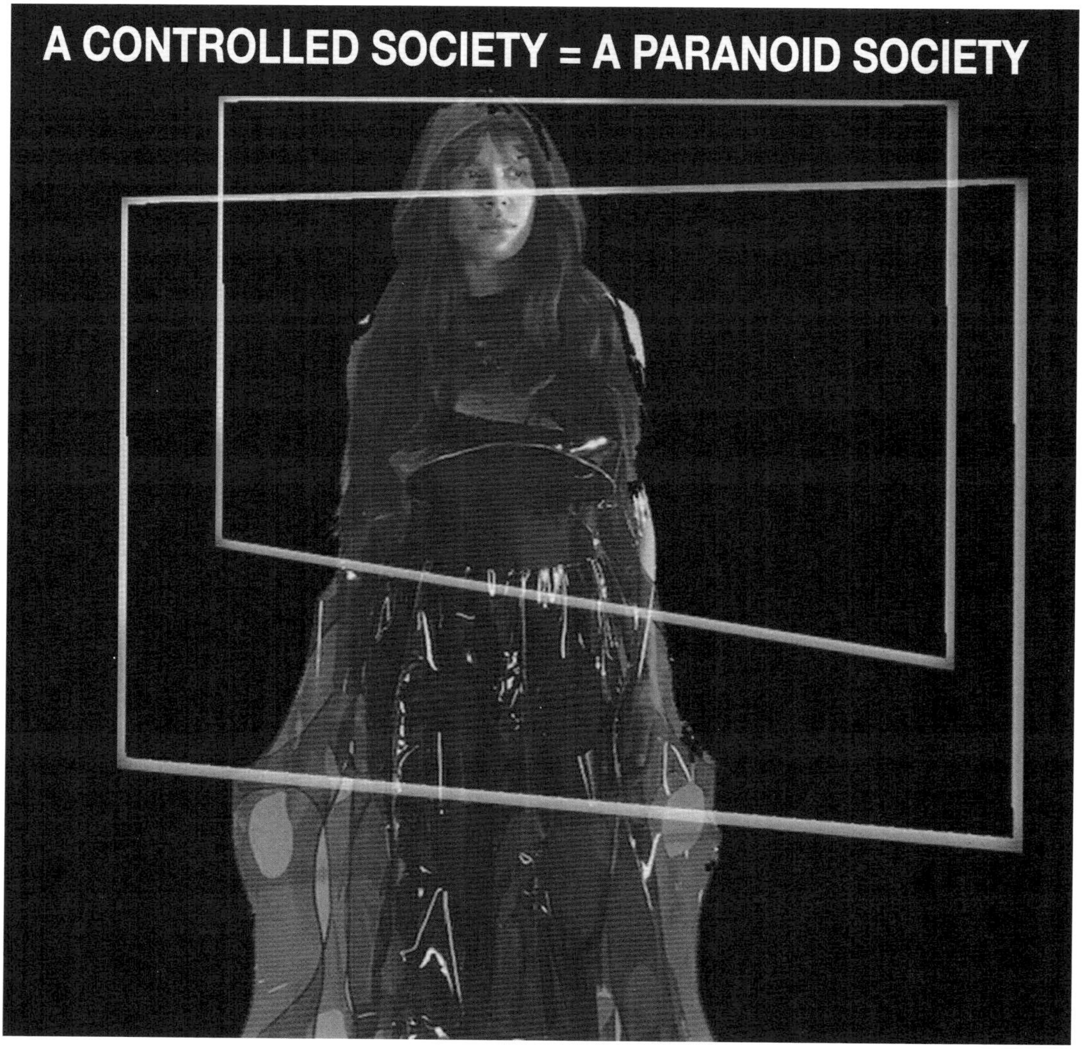

Lynn Hershman Leeson, *Shadow Stalker*, 2019. Still from the film/installation

thought it was fiction, and the more people said they related to it. I think part of intimacy is getting rid of artifice and going into the deepest part of your soul to express it in a simple way. We all suffer and we've all been trained not to talk about it or share it.

SO: A beautiful thing happens when the personal becomes universal. Were you actively trying to tap into this universality, or did it become apparent through viewer responses?

LHL: I never thought I would show the diaries, so I was very surprised that they were distributed and even translated into five languages. It shocked me that they reached so many people. As artists, we express the world or ourselves in the world at a particular moment. The work is like a fingerprint—it doesn't change that much. We see things from our own perspective and filter it.

SO: Throughout your career you've had your finger on the pulse, working with emerging technologies and collaborating with leading scientists and technologists. What is your process for choosing your next medium? In the case of *The Electronic Diaries,* it was a matter of learning how to use a new material. Does the process of learning how to use a new technology often prompt ideas for artworks?

LHL: [The artworks] all come from the story or the idea, and then I either use technology that exists or invent new technology for the project. We [the artist and her collaborators] need to find a way to manifest the idea, a way of believing in the impossible and in inventing our own reality. With [the interactive touchscreen videodisc] *Deep Contact* (1984), touchscreens weren't available, so we had to make our own. Working in this way is kind of like hijacking a car: you experiment and try plugging in different things until you get it to work. Unfortunately, you then drive around for a mile and find that somebody else has already invented a better one.

SO: Well, you're using the touchscreen technology for the very specific function: creating an artwork. The invention doesn't need to be used in the wider world. You didn't create the touchscreen, for example, to be used in a smartphone, but rather to address a specific set of theoretical questions: Can we draw a line between digital and physical bodies? What is the emotional experience of being a digital voyeur? What might intimacy look like in the digital age?

LHL: Exactly. The same thing happened when I made *Agent Ruby* (1998–2002), the AI chatbot. I created it in 1998, so pre-*Teknolust* (2002) [a sci-fi feature film about a bio-geneticist who creates cyborg clones with her own DNA] but very much a part of that thinking. I had eighteen programmers from around the world who responded to an open-source question that was eventually adapted by

Colin Klingman, who was Ruby's ultimate programmer. The other programmers all wanted to find out how to make a computer talk back and would suggest useful ideas, but they were not on the team that saw the project through. Of course, it took us years, and then [computer scientist] Richard Wallace figured it out right before and presented a show near us.

What we did with *Agent Ruby* and *Dina* (2000–2006) [an AI web bot] is what Siri went on to do twelve years later. I never think of making things as profitable vehicles, but rather to find out if the project is even possible.

SO: There's also then the issue of communication and access. You had these very open and collaborative teams, whereas I'd imagine some of the technology spaces might have been closed off or more competitive, for fear of leaking trade secrets and new knowledge.

LHL: It's true. While working on the AI chatbots, someone loaned us software to use, and the guy overseeing the technology kept coming over and watching what we were doing. Finally, my programmer, who is more savvy in this world than I am, said, "Don't tell them anything." I think this information was used to develop Siri.

SO: I imagine your research was very fruitful for your programmer collaborators as well as the alleged spies, considering the amount of information that *Ruby* was being fed with. She's a web-based AI chatbot, and I understand that she "remembers" or stores users' questions, and is able to recognize past users' voices, moods, and emotions. Is she still collecting information from users, or has her domain been closed?

LHL: She's been collecting information and having chats since 2001. When I showed her in 2013 [at SFMOMA], I learned that things on the internet never die. At that time, she had information from people who wrote to her from all over the world. She became a global portrait of people's thoughts—whether about 9/11 or Bush's election or Obama's election. She's now in the permanent collection at SFMOMA.

SO: While *Ruby* exists purely as a digital avatar, Roberta (1973–96) exists in the digital realm but also in the physical world, both through your body and through those of actors.[2] How does Roberta's identity align with yours? Do you see her as an extension of you, or do you understand her as a discrete being whose existence you're only helping to facilitate?

LHL: I tried to make Roberta her own person with her own background, language, gait, psychiatric record, pharmaceuticals, and so forth. But we did share a heartbeat, so as hard as I tried to make her

2 Roberta Breitmore was a fictional character that Hershman Leeson originally played herself. Roberta first forayed into the world when checking into the Dante Hotel in San Francisco in 1973. She subsequently opened a bank account, saw a therapist, rented an apartment, and even communicated with potential roommates that responded to her ad in a local paper. After four years, Hershman Leeson paid a series of actors to play Roberta. Sometimes, multiple Robertas appeared at the same gallery opening. The performance—which was documented in drawings, photographs, and via personal records (a driver's license, credit cards)—ended in 1978 when Hershman Leeson had Roberta exorcised. Hershman Leeson later commissioned Zap Comix artist Spain Rodriguez to archive Roberta's life in a graphic novel. A version of Roberta was revived as *CyberRoberta* (1996), a doll with camera-eyes who is outfitted with Roberta's signature blonde hair, sunglasses, and polka-dot dress.

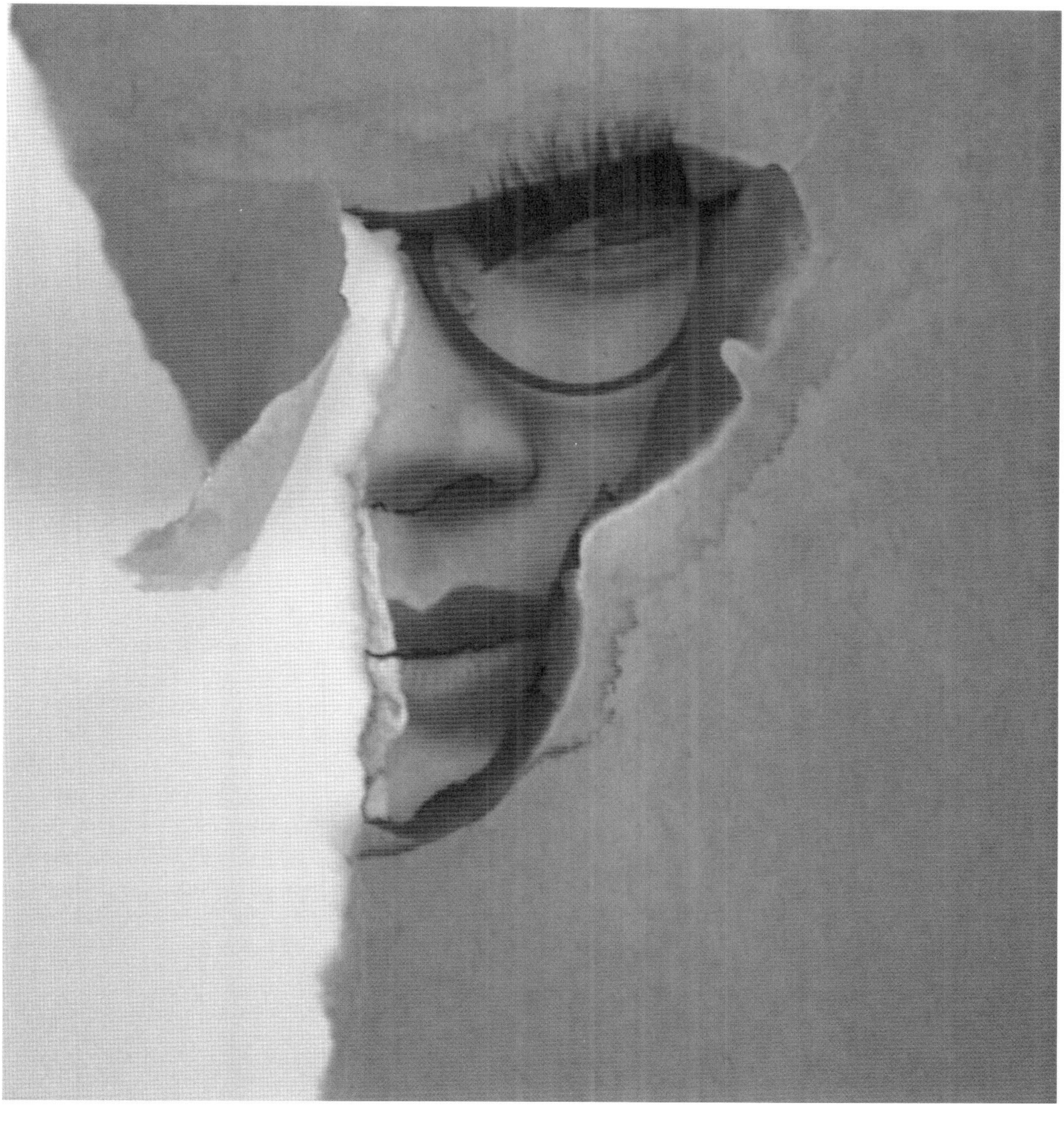

Lynn Hershman Leeson, drawing/collage, 2020

separate, to make her as an archetype of what was going on at the time, we still overlapped. I had to do the multiples in order to test whether or not she was reflecting me or the wider culture. They were camouflaged experiments that went out into the world and created a profile of what it was like in those days to live as a young, blonde, single woman.

SO: Did you feel there was ever a power struggle between you and Roberta over who had claims to the body?

LHL: Only in her psychiatric sessions. I didn't have enough money for both of us to go, and I wanted to use that time for myself. This was the only conflict though; otherwise, she was pretty specific and distinct from me.

SO: As two people who shared one body, how did you communicate?

LHL: I've wondered this in relation to conjoined twins. Who controls what? What kind of conflicts do they have?

SO: Right, which—the psychological, the biological, or the digital body—carries more weight? Do you privilege any over the others?

LHL: In the roughly twenty years since discovering the human genome, I think that the idea of a double helix applies to how we relate to our physical and virtual worlds. We are constantly moving between both, and I think we need both to become whole and fully human. We can't really be physically human without the virtual; however, I do think that technology is getting the upper hand in everything we do.

SO: Do you think that we fetishize technology too much, or just that we're now increasingly reliant on it?

LHL: I think that there's a loneliness entrenched in the very essence of being human that we often try to avoid. This feeling of loss, or that we are missing something, is only exacerbated by the fact that information is moving really fast. However, I also think that people can become more intimate online then in person.

SO: As many of us have been in various stages of physical isolation in our homes for months now, this sense of loneliness is increasingly coupled with a struggle to define what counts as "reality." I've been feeling more disconnected from the world outside my apartment, which is now wholly mediated by screens. As someone who has been actively working within the virtual space for years, have you noticed a greater public fluency with digital technologies that cater to forms of virtual intimacy? I'm thinking about chatbots and forums, as well as the rise in video communications software such as Zoom.

LHL: I think what has changed is the power of connectivity: our digital selves have become so influential. People can be surveilled through nearly invisible programming en masse. The idea with Roberta was that she was the archetype of a person at a specific time in history [the 1970s]. Today, there are invisible algorithms that track and predict our identities and have records of our digital files. Now, it seems so primitive to have just a camera as a surveillance tool. When things are more sophisticated and you don't have access to how they work, that becomes really frightening—especially when it's happening on a global level and affecting so many people in this invisible, toxic way.

SO: The amount of information that we give up every day—willingly or unknowingly—is concerning . . . That reminds me, do you know the artist Heather Dewey-Hagborg?

LHL: Yes.

SO: Your comment about omnipresent surveillance systems reminds me of her project *Stranger Visions* (2012–13), where she collected trash that would likely have remnants of DNA from New York City streets—cigarette butts, chewing gum, loose hair, etc.—and used facial-modeling algorithms to create speculative portraits with the genetic information that she found. The work brings up issues around anonymity and consent.

LHL: With [Dewey-Hagborg's] project, it is important to stress that the portraits aren't real. The algorithm takes liberties in constructing an image, as DNA alone can't definitively create a profile for a person.

SO: While the twentieth century was celebrated for advancements in physics, the twenty-first century has taken a move toward biology. Considering this shift and all of the threats it poses to one's control over their own body, I'm curious how much biological agency you think we have as citizens today.

LHL: I think we have some agency; however, mapping the genome was such a dynamic force that affected every living thing as well as our understanding of hybridity.

SO: Do we have the tools or language to push back on biologically focused surveillance apparatuses and advancements in biotechnology that lend themselves to agricultural manipulation among other things?

LHL: We are at a point in time where we really can't accept things as they are. By looking the other way for so long, we have destroyed the

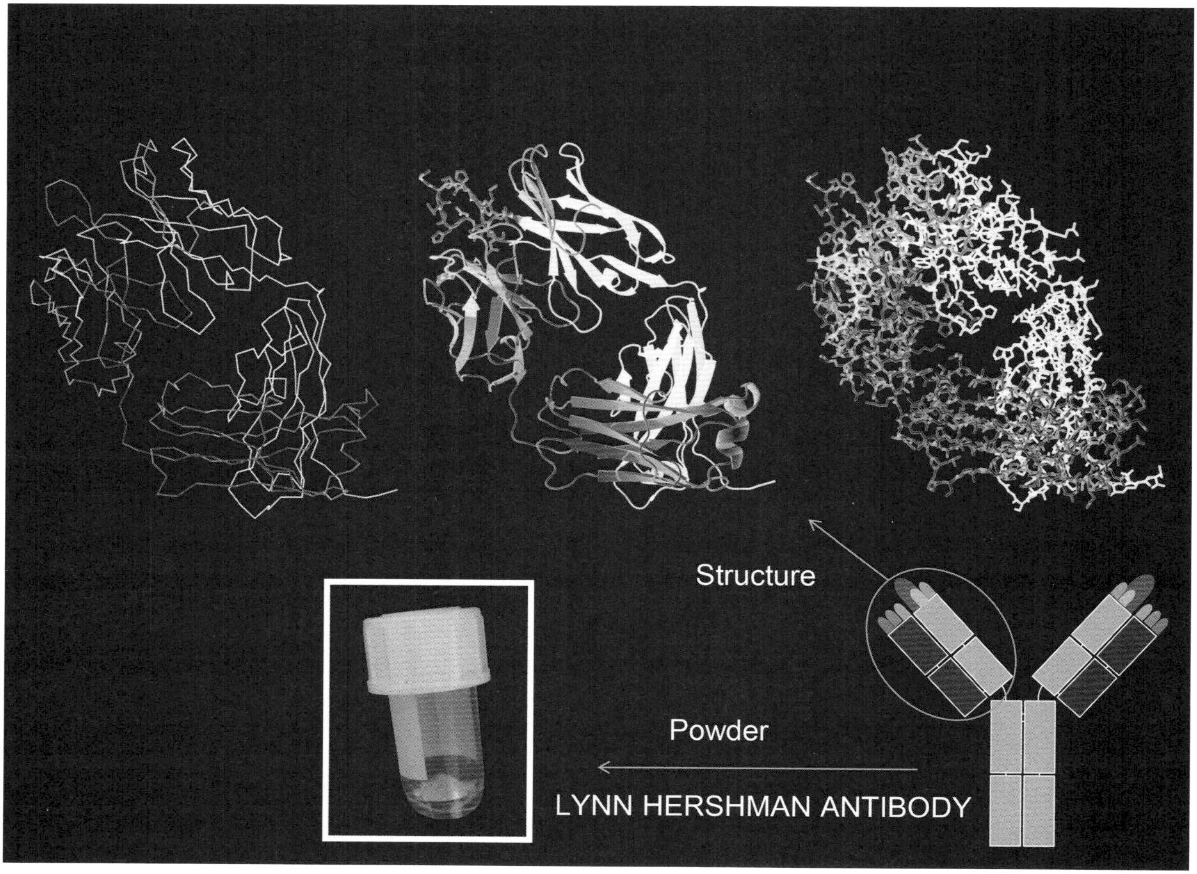

Lynn Hershman Leeson, design for the creation of the *LYNN HERSHMAN Antibody*, with Dr. Thomas Huber

planet, and now we don't have a choice. We've spoken before about how we are currently going through a period of corrections.[3] I think there are strategies to shift the power away from megalomaniacal capitalists who are exploiting most of the world, toward generosity and new systems of sharing. I've been thinking about Anna Lowenhaupt Tsing's *The Mushroom at the End of the World: On the Possibility of Life in Capitalist Ruins* (2015), where she suggests looking to tree root systems as models for how we should adapt our communication methods to survive.

SO: Right, we need to consider ecological strategies and even shift the conversation away from a human-centric focus, ideally more toward a planetary and interspecies approach. What really struck me from her book was her imperative to reconsider time itself: she argues that "conventionally, historians look only at human remainders, such as archives and diaries, but there is no reason not to spread our attention to the tracks and traces of nonhumans, as these contribute to our common landscapes."[4] We agree that there's an urgent need for new strategies for collaboration and making. But is your work optimistic for the future?

LHL: I like to think that most of the work I do is hopeful and has a base element of utopian thought. I want people to walk away thinking they can do something, that they don't have to accept the situation as it exists. I want to use art and death to allow people to be more aware of their power and their ability to make choices in their own lives.

3 Lynn Hershman Leeson, interview by Samantha Ozer, *Artforum*, June 29, 2020, https://www.artforum.com/interviews/the-artist-talks-about-her-work-with-antibodies-83458.

4 Anna Lowenhaupt Tsing, *The Mushroom at the End of the World: On the Possibility of Life in Capitalist Ruins* (Princeton: Princeton University Press, 2015), 168.

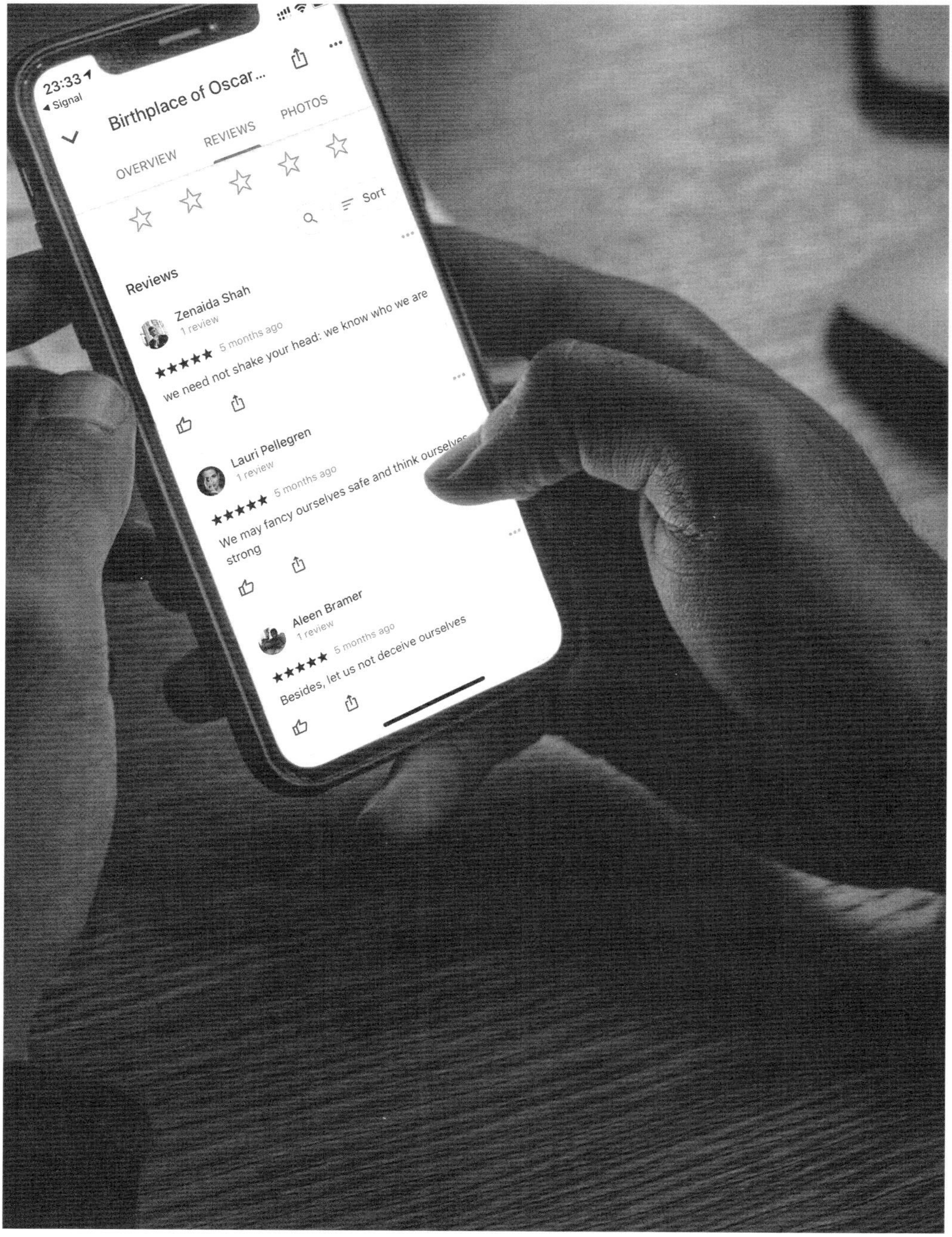

Constant Dullaart, *This Unjust Mirror*. https://soundcloud.com/kanyewest/real-friends. https://www.google.com/maps/place/Birthplace+of+Oscar+Wilde/@53.342725,-6.2527111,17z/data=!3m1!4b1!4m7!3m6!1s0x48670f68a9098e61:0x40d59b5375922d05!8m2!3d53.342725!4d-6.2505224!9m1!1b1. https://www.youtube.com/watch?v=aPrvkbz9KT4

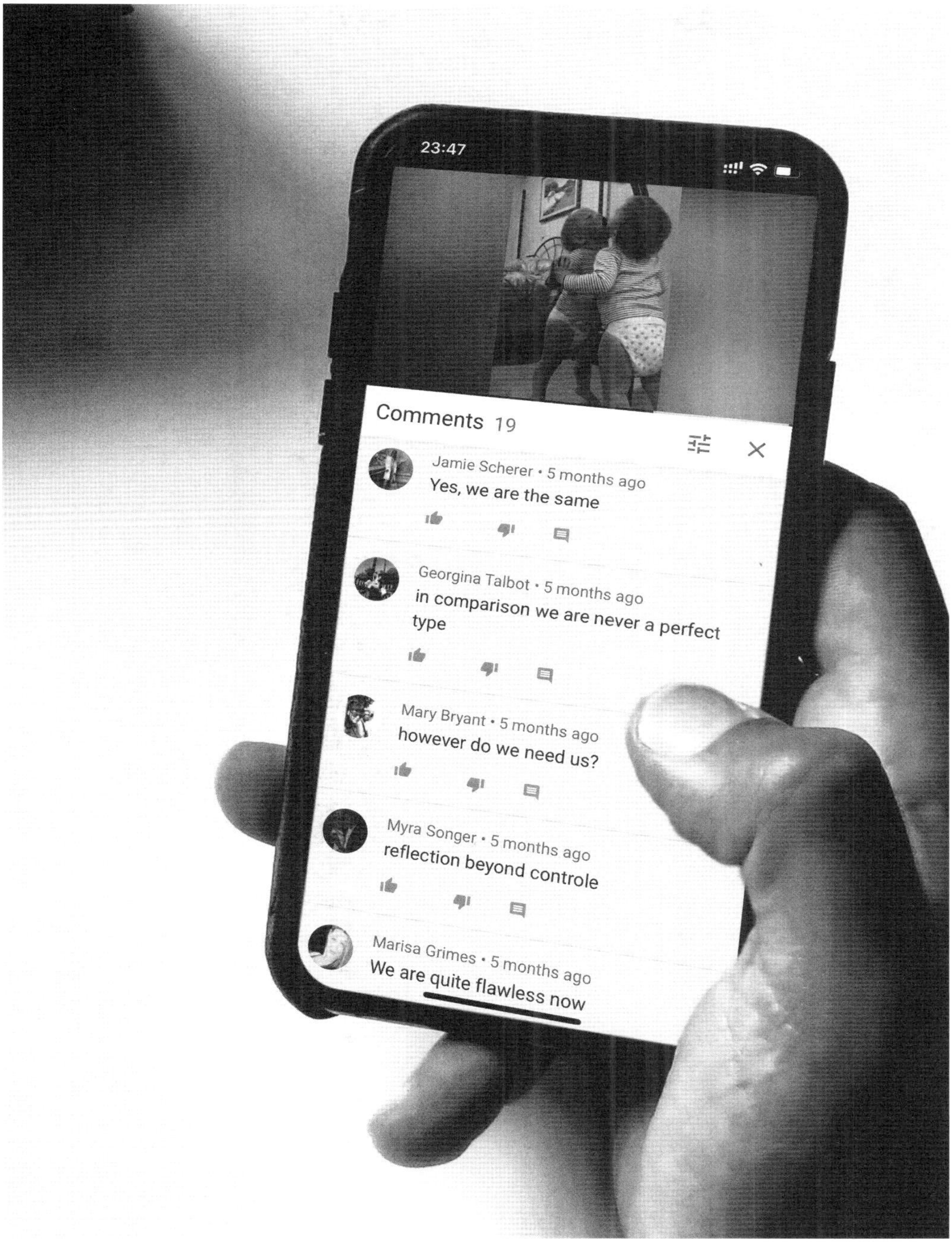
23:47
Comments 19
Jamie Scherer • 5 months ago
Yes, we are the same
Georgina Talbot • 5 months ago
in comparison we are never a perfect type
Mary Bryant • 5 months ago
however do we need us?
Myra Songer • 5 months ago
reflection beyond controle
Marisa Grimes • 5 months ago
We are quite flawless now

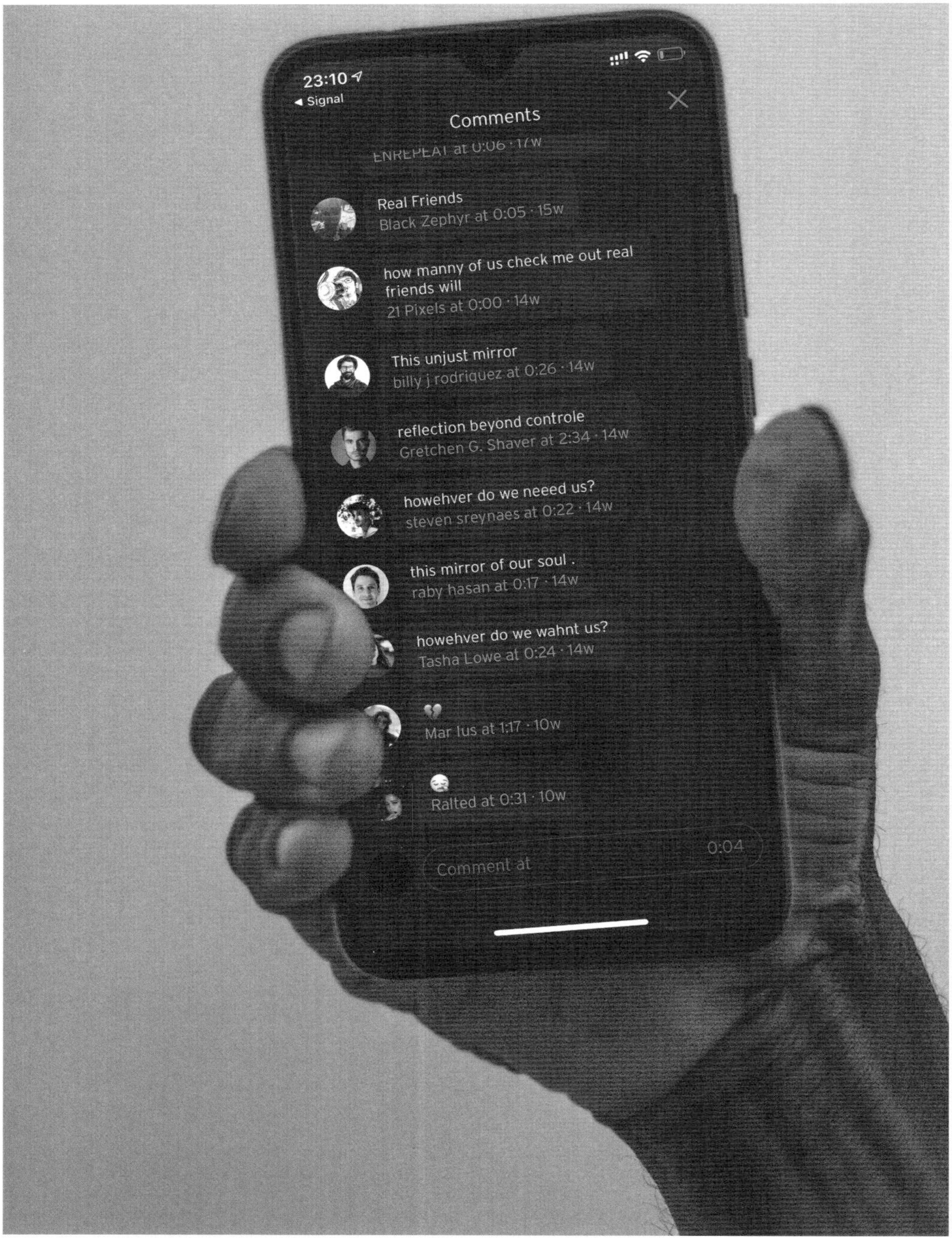
23:10
◂ Signal
Comments
Real Friends
Black Zephyr at 0:05 · 15w
how manny of us check me out real friends will
21 Pixels at 0:00 · 14w
This unjust mirror
billy j rodriquez at 0:26 · 14w
reflection beyond controle
Gretchen G. Shaver at 2:34 · 14w
howehver do we neeed us?
steven sreynaes at 0:22 · 14w
this mirror of our soul .
raby hasan at 0:17 · 14w
howehver do we wahnt us?
Tasha Lowe at 0:24 · 14w
Mar lus at 1:17 · 10w
Ralted at 0:31 · 10w
Comment at
0:04

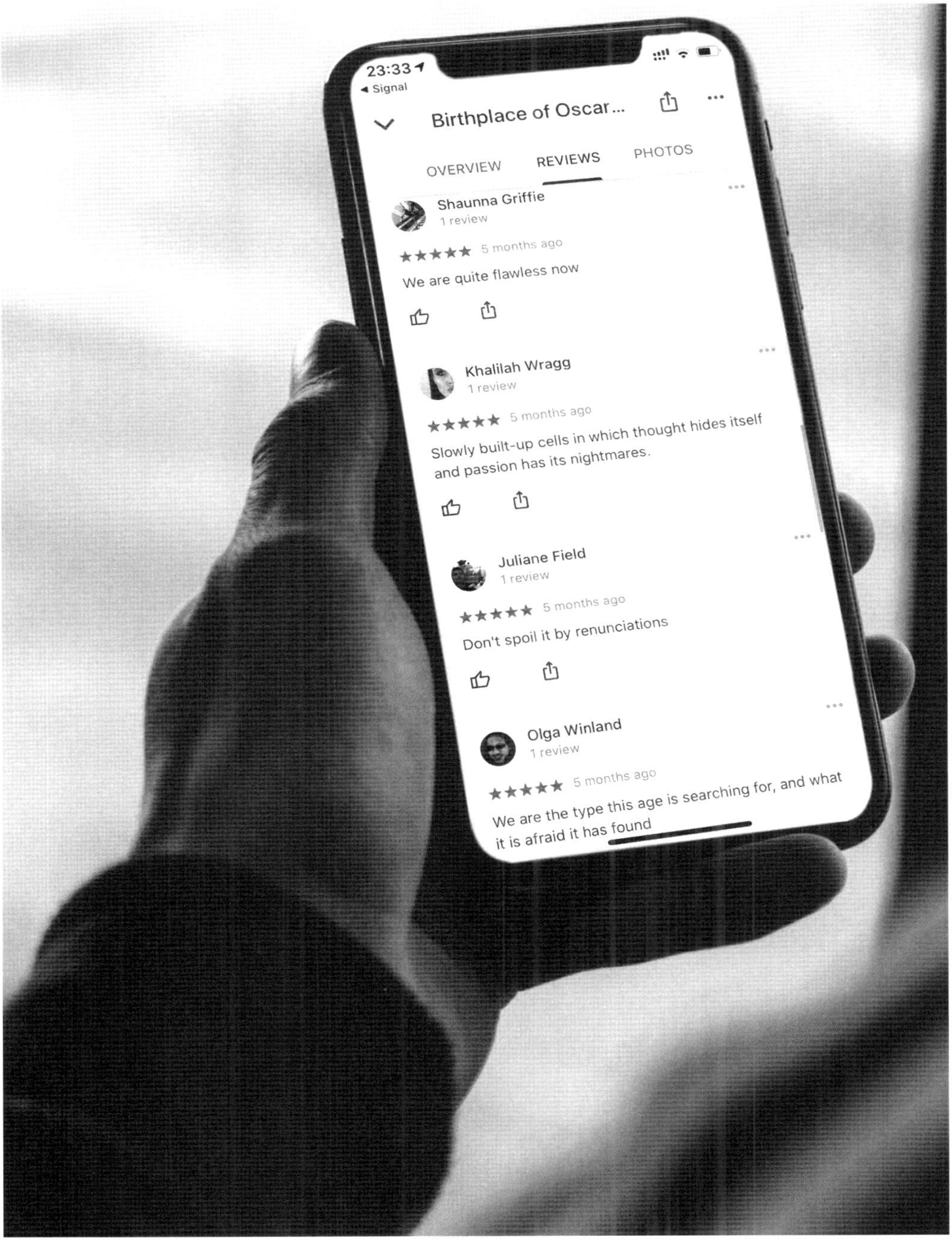
23:33
Signal
Birthplace of Oscar...
OVERVIEW
REVIEWS
PHOTOS
Shaunna Griffie
1 review
5 months ago
We are quite flawless now
Khalilah Wragg
1 review
5 months ago
Slowly built-up cells in which thought hides itself and passion has its nightmares.
Juliane Field
1 review
5 months ago
Don't spoil it by renunciations
Olga Winland
1 review
5 months ago
We are the type this age is searching for, and what it is afraid it has found

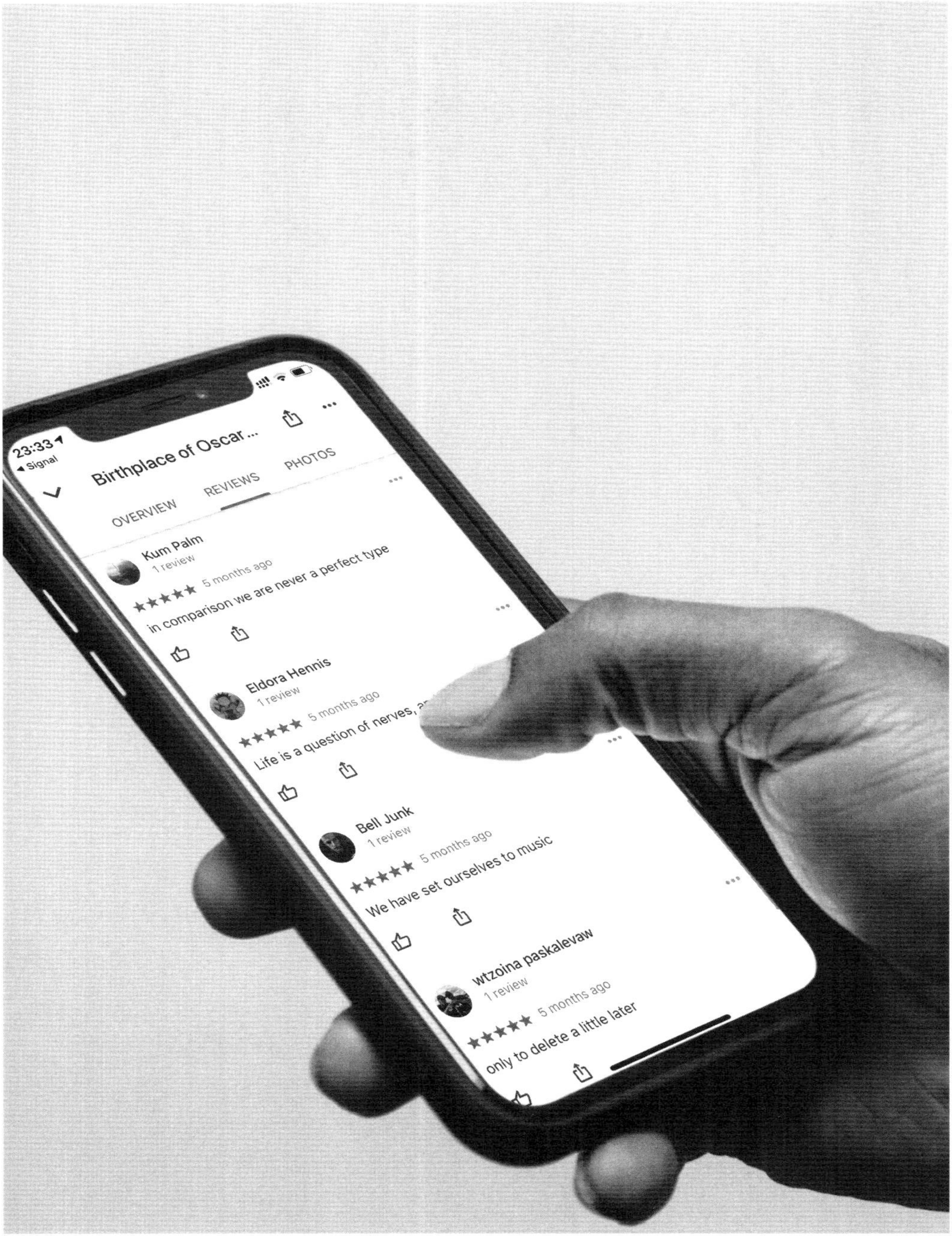
23:33
◂ Signal
Birthplace of Oscar ...
OVERVIEW
REVIEWS
PHOTOS
Kum Palm
1 review
★★★★★ 5 months ago
in comparison we are never a perfect type
Eldora Hennis
1 review
★★★★★ 5 months ago
Life is a question of nerves,
Bell Junk
1 review
★★★★★ 5 months ago
We have set ourselves to music
wtzoina paskalevaw
1 review
★★★★★ 5 months ago
only to delete a little later

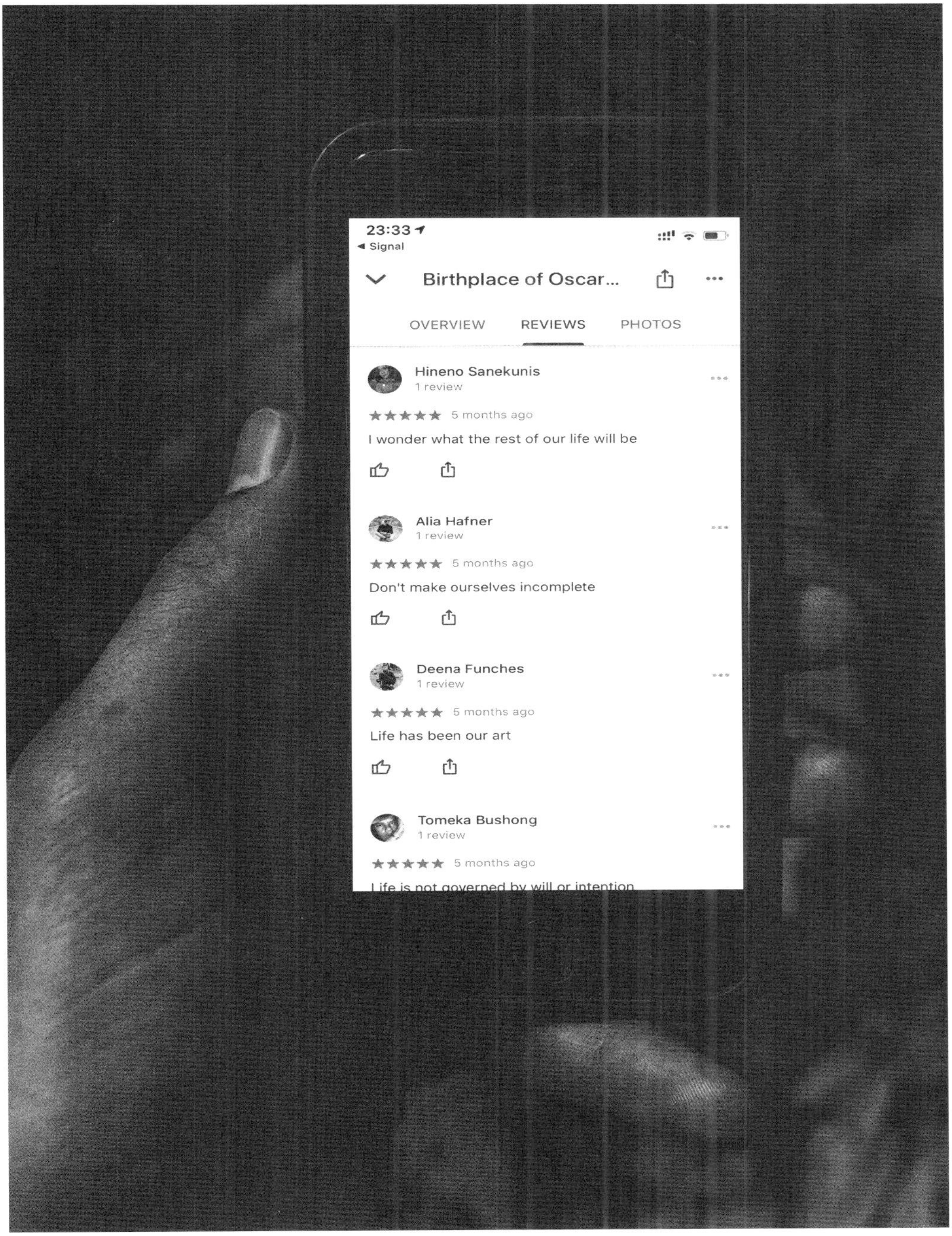
23:33
◂ Signal
Birthplace of Oscar...
OVERVIEW
REVIEWS
PHOTOS
Hineno Sanekunis
1 review
5 months ago
I wonder what the rest of our life will be
Alia Hafner
1 review
5 months ago
Don't make ourselves incomplete
Deena Funches
1 review
5 months ago
Life has been our art
Tomeka Bushong
1 review
5 months ago
Life is not governed by will or intention

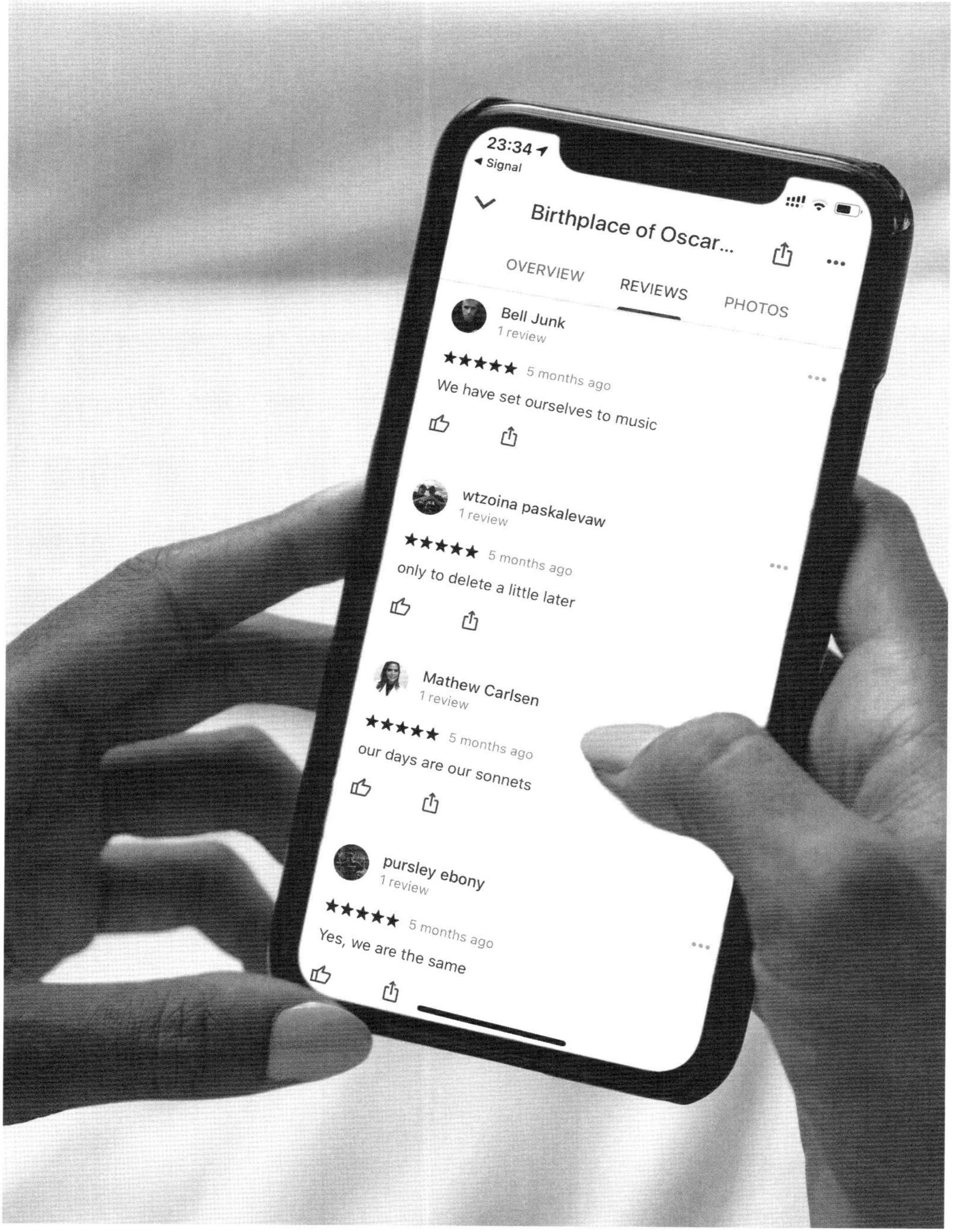
23:34
◂ Signal
Birthplace of Oscar...
OVERVIEW
REVIEWS
PHOTOS
Bell Junk
1 review
5 months ago
We have set ourselves to music
wtzoina paskalevaw
1 review
5 months ago
only to delete a little later
Mathew Carlsen
1 review
5 months ago
our days are our sonnets
pursley ebony
1 review
5 months ago
Yes, we are the same

A social attitude vis-à-vis machines [. . .] expresses [. . .] the position of desire in the infrastructure.[1]

This text maps a constellation of trajectories oscillating between intimacy and the web of machinic infrastructure in which intimacy operates. In doing so, it critically reads intimacy as drawing on a space of interiority contingent upon this machinery. In other words, rather than suspending intimacy, this text probes an alternative notion: *extimacy*. In psychoanalysis, this term addresses the subject's immanent relation with radical exteriority. Drawing on Félix Guattari's psychiatric experiments with media such as voice recorders and film, and his critique of Lacan's structuralist idea of the subject, this text explores the entanglement between the *unconscious* and the *machine*. Extimacy is no longer understood as a paradoxical interplay between the subject and its innermost exteriority. The question is rather: How does the problem of the topology of the psyche elucidate our present machinic condition,[2] both in its historical development as well as in its present digital configuration?

Interior Machina

In 1959, Jacques Lacan coined the term *extimité*, critically addressing the problem of the topology of the psyche. He could hardly have imagined the profound transformations of the half century to come: a world in which geographical distances so radically changed their meaning—the world of the technical colonization of the unconscious.[3] However, this same new world was also inhabited by a multiplicity of new senses, by desires inflamed by a refusal to accept as given or natural the regime of sexual difference; a world of transcultural global communities formed and disseminated in opposition to the territorialized state politics. Yet precisely these then distant constellations incite us to reevaluate Lacan's concept of extimacy.

Lacan's inversion of "intimacy" through the prefix "ex" might appear paradoxical at first glance, as it points at something *outside* located simultaneously *inside*. The French *intimité* (intimacy) designates not only closeness and proximity but refers also to a topological quality of the deepest, innermost part of one's being: a locus, *intimateness*. Such topological intensification is also reflected in the

Unknown, Illustration of the Camera Obscura Principle, 18th century

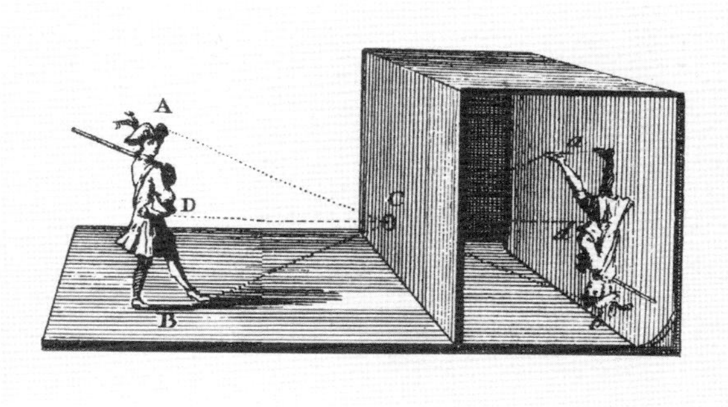

1 Félix Guattari and Gilles Deleuze, "Balance-Sheet for 'Desiring-Machines,'" in Guattari, *Chaosophy: Texts and Interviews 1972–1977*, ed. Sylvère Lotringer, trans. David L Sweet, Jarred Becker, and Taylor Adkins, Semiotext(e) Foreign Agents Series (Los Angeles: Semiotext(e), 2009), 114.

2 Machinic: "of or relating to machines." Wiktionary, s.v. "Machinic," accessed March 16, 2019, https://en.wiktionary.org/wiki/machinic. Guattari and Deleuze use this term throughout their work to describe the machinic qualities of reality, which this essay explicates below.

3 Describing the new ethology in the age of the digital, conceptualized as "new feelings," Alexandra Molotkow refers to the vertiginous immediacy of relations and potential contacts in social networks. Molotkow, "New Feelings: Crush Fatigue," *Real Life*, August 1, 2018, https://reallifemag.com/new-feelings-crush-fatigue.

4 Jacques-Alain Miller, "Extimité," *Prose Studies* 11, no. 3 (December 1988): 122.

5 Sigmund Freud, note of August 22, 1938, in Freud, *Standard Edition of the Complete Psychological Works*, trans. James Strachey (London: Hogarth Press, 1964), 23:300, quoted in Edward S. Casey, *The Fate of Place: A Philosophical History* (Berkeley: University of California Press, 1997), 456n6.

6 Miller, "Extimité," 123.

7 Lacan not only served as Pablo Picasso's therapist but also published texts in Surrealist journal *Minotaure*.

8 Henning Schmidgen, *Das Unbewusste der Maschinen: Konzeptionen des Psychischen bei Guattari, Deleuze und Lacan* (Munich: Wilhem Fink Verlag, 1997), 114–115; see also Svitlana Matviyenko, "Lacan's Cybernetics" (PhD diss., University of Western Ontario, 2015), https://ir.lib.uwo.ca/etd/3461; on the influences leading from cybernetics to French theory more

etymology of the word, since the Latin *intimus* is a superlative of "interior." Intimacy marks "an effort on the part of language to reach the deepest point in the interior."[4]

Speaking of extimacy at decisive junctures, Lacan took up a fundamental problem that obsessed Sigmund Freud. In one of Freud's final, and rather enigmatic, notes, he envisioned the psyche as being radically extended into space:

> Space may be the projection of the extension of the psychical apparatus. No other derivation is probable. Instead of Kant's *a priori* determinants of our psychical apparatus. Psyche is extended; knows nothing about it.[5]

Freud counters Immanuel Kant's major philosophical concept: time and space as *a priori* conditions for psychic processes and cognition. Following his path, Lacan later derived that the psyche is *extimate* to itself, *is* unconscious. Extimacy is not the opposite of intimacy; "Extimacy says that the intimate is Other—like a foreign body, a parasite."[6] The Other, the foreign intimate, the unconscious: these are synonyms, figures ruthlessly driven out of themselves.

Observing symptoms of paranoid psychosis, Lacan described behavior escaping patients' will as "automatic phenomena." Such acts mediated something imposed from the outside. In the 1930s Lacan was close with Surrealists who practiced *écriture automatique*, the technique of giving immediate expression to what André Breton called "inner discourse."[7] What Freud conceived as *Wiederholungszwang* or "repetition compulsion" in *Beyond the Pleasure Principle*, Lacan conceptualized as *automatisme de répétition*. Deliberately foregrounding such a machinic quality of the psyche, Lacan insists on the specific function of the unconscious as memory of the symbolic. Later, in his famous 1956 cybernetic seminar *The Seminar on The Purloined Letter*, Lacan transposed these spectral automatisms into a language shaped by contemporary information theory:[8]

> For we have learned to conceive of the signifier as sustaining itself only in a displacement comparable to that found in electric news strips or in the rotating memories of our machines-that-think-like-men, this because of the alternating operation which is its principle, requiring it to leave its place, even though it returns to it by a circular path.[9]

In his translucent and groundbreaking study on Guattari's machinic unconscious, *Das Unbewusste der Maschinen* (The unconscious of machines, 1997), Henning Schmidgen analyzes Lacan's association of psychic processes and information technology.[10] He also points out Lacan's extension of Freud's

generally, see Bernard Dionysius Geoghegan, "From Information Theory to French Theory: Jakobson, Lévi-Strauss, and the Cybernetic Apparatus," *Critical Inquiry* 38, no. 1 (September 2011): 96–126.

9 Jacques Lacan, "Seminar on 'The Purloined Letter,'" in *Ecrits: The First Complete Edition in English*, trans. Bruce Fink (New York: W.W. Norton & Co, 2006), 21. Also available at https://www.lacan.com/purloined.htm#11x.

10 Excerpts of which are available in English: Henning Schmidgen, "Machinic Subjectivity: Psychoanalysis, Cybernetics, and the Assemblage of Technology," trans. Gloria Custance (self pub., 2019).

contentions. In Lacan's words, "It is not only the subject, but the subjects, grasped in their intersubjectivity, who line up [. . .] and who [. . .] model their very being on the moment of the signifying chain which traverses them."[11] For Lacan, such machinic procedures encapsulate the core of intersubjective relations. Lacan's equation places the unconscious on the level of the machine. Or more precisely, intersubjectivity is a machine determined by the chained, metric displacement of the signifier. Coupling the alterity of the unconscious with the machine, Lacan undertakes a radical step toward dehumanizing the psyche, exteriorizing it, and transposing the interfering function from the subject on the machine. Yet by the same gesture, he makes the subject a prisoner of its "translucent, aseptic, imputrescible and timeless signifying chains."[12]

In his 1966 essay "From One Sign to the Other," published in *Psychoanalysis and Transversality* (1972), Guattari offers a scathing account of Lacan's model of the unconscious:

> The 'I' defines nothing. [. . .] Grasped by the other on the edge of myself I ebb to the other end of the chain. The coordinates cross. The I for I was only a possible mirage in the intimacy of the other for me. But this other, to the extent that it refers to an other for the other, from which any orientation is impossible, forbids me any security related to the true position of the subject and leaves me beset by the vertigo of an eternal return to my first effusions with death.[13]

In this "infinite game of references from one structure to another," Lacan's subject as well as the machine correspond with the totalizing set of the signifier.[14] Guattari uncovers the tautological premise of such a model, a model promoting an omnipotent agent to which all possible signification is subordinated.

From One Machine to Another

While Guattari excoriated Lacan for modeling the psyche on a totalizing and reductive conception of the machine, his own theoretical outlook by no means repudiated machinic concepts. Rather the opposite: Guattari would develop a veritably machinic theory—yet this move went hand-in-hand with a radical reorientation of the concept of the machine itself, largely derived from his psychotherapeutic experience. Henceforth, the machine and the machinic would refer to productive interconnections between heterogeneous domains and elements, encompassing the social, psychical, political, and aesthetic just as much as the technical.

This new concept took center stage in *Anti-Oedipus* (1973), the famous manifesto on desiring

11 Lacan, "Seminar on 'The Purloined Letter,'" 21.

12 Félix Guattari, "From One Sign to the Other (Excerpts)," in *Psychoanalysis and Transversality: Texts and Interviews 1955–1971*, trans. Ames Hodges (Los Angeles: Semiotext(e), 2015), 200.

13 Guattari, "From One Sign to the Other," 200–201.

14 Guattari, "From One Sign to the Other," 199.

15 Gilles Deleuze and Félix Guattari, *Anti-Oedipus: Capitalism and Schizophrenia*, trans. Robert Hurley, Mark Seem, and Helen R. Lane (Minneapolis: University of Minnesota Press, 1983), 1.

16 La Borde is situated two hundred kilometers south of Paris in Cour-Cheverny; Guattari was invited by its director, the psychiatrist Jean Oury. While there, Guattari also attended Lacan's lectures in Paris and shared the knowledge with the team at La Borde. An elaborated account of Guattari's activity in La Borde is given in Schmidgen, *Das Unbewusste der Maschinen*, as well as in Henning Schmidgen, ed., *Die Guattari-Tapes: Gespräche mit Antonio Negri, Jean Oury, Jean-Claude Polack, Élisabeth Roudinesco, Danielle Sivadon und Paul Virilio* (Berlin: Merve, 2019).

17 Félix Guattari, "La Borde: A Clinic Unlike Any Other," in *Chaosophy: Texts and Interviews 1972–1977*, ed. Sylvère Lotringer, trans. David L Sweet, Jarred Becker, and Taylor Adkins, Semiotext(e) Foreign Agents Series (Los Angeles: Semiotext(e), 2009), 178.

18 Guattari, "La Borde," 179.

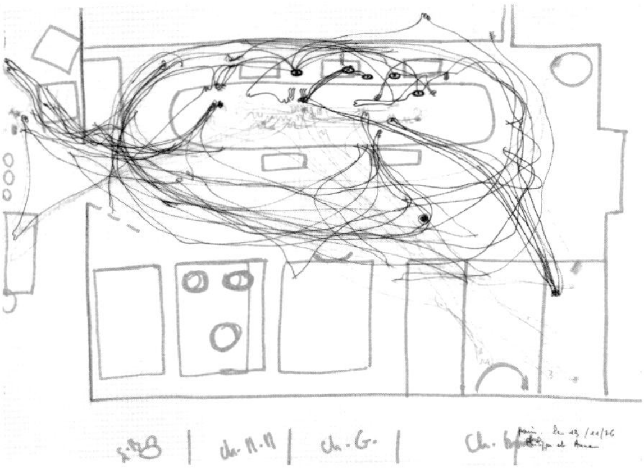

Fernand Deligny's *Network in Cévennes*. Trajectories or "wander lines" of Philippe and Anne (two autistic children) in the kitchen of the house Y in Monoblet, November 13, 1976. Traced by Thierry Bazzana et Marie-Madeleine Godet

machines written with Gilles Deleuze. Both its criticism of psychoanalysis as well as its rampant machinism drew on Guattari's fifteen years of therapeutic practice. While machinic vocabulary was merely metaphorical for Lacan, Guattari elaborated and implemented a theory of "real [...] not figurative" machines that applied to socio-institutional contexts as well as the actual media and machines in his therapeutic practice.[15] Two formative episodes elucidate this shift, and this essay will extrapolate from these to present-day assemblages of the social, psychical, and technical.

First, when Guattari joined the psychiatric clinic of La Borde, he was responsible for implementing "institutional psychotherapy," a therapy based on group methods, workshops, and the critical restructuring of the institution, as well as a series of media and artistic practices operated together with the patients.[16] Guattari describes the new ergonomic structure of La Borde as an "internal mini-revolution": a rotational principle required "all service personnel work to be integrated with medical work, and that, reciprocally, medical staff be drafted for material tasks such as cleaning, cooking, dishwashing, maintenance."[17] As Guattari recalls, "an old washerwoman proved very capable at running the print workshop and editorial committee of the newspaper; another excelled in sporting activities, a former metallurgist showed great talent leading mime shows."[18] This principle aimed at a gradual desegregation and disalienation

René Laloux, *Les dents du singe*, 1961. Animation, 5:34

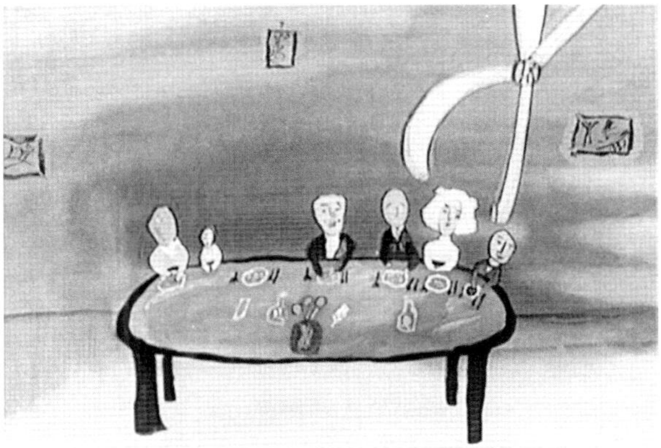

of the institution's internal relationships—the patient-doctor relationship as well as that between medical staff and service personnel. A collectively devised grid that formalized and diagrammed the division of labor constituted this process's fulcrum, mediating between tasks and subjectivities.[19] Against the ossified prior routine, this reorganization of labor installed a machine intersecting the clinic's institutional, social, and psychical domains that intensified the heterogeneity of activities and contacts within the milieu.

"The institutional machine that we positioned didn't simply remodel the existing subjectivities, but endeavored, instead, to produce a new type of subjectivity."[20] Such an idea of disalienation implied—besides the manifest positioning against monotonous "seriality" and hierarchy of institutions—a dismantling of another alterity: the alienation of the unconscious through Lacan's Other. Guattari's "institutional analysis" opened the Pandora's box of the unconscious toward social and political dimensions, such as "health, pedagogy, prison conditions, femininity, architecture, urbanism."[21] Furthermore, machinic interventions opened the rigid mirror corridor of Lacan's imaginary through an active use of imagination: staging of theater plays, production of exhibitions (such as "Le siècle de Kafka" at the Centre Georges Pompidou, 1984), films, and journals served as "projective equivalents," therapeutic techniques helping to restore patients' bodies and minds.

Second, for Guattari, machines were equivalent to psyches on the basis of their active intervention, or even interaction: from one machine to another, from psyche to machine. This elucidates his formative experience with the case of R.A., a patient showing symptoms of psychosis who refused any constructive interaction. In his therapeutic cure with the patient Guattari used a tape recorder.[22] This machine *par excellence* broke the two-person psychotherapeutic structure by introducing a crucial third instance: "The recorded voice, the drawling tone, the hesitations, the breaks, the constant incoherence revolted him, and he took me as a witness that he must have truly fallen 'lower than everything' to end up speaking like that."[23]

Guattari also mentions an amateur film made at La Borde: when R.A. watched the recording, he—to Guattari's surprise—recognized himself participating in various activities. Next, Guattari asked R.A. to transcribe Franz Kafka's novel *The Castle* (1926), having previously noted similarities between the patient and the author's psychopathological structure and external appearance. This projective exercise gradually enabled the patient not only to speak but also to keep his own diary, fragments of which feature in *Psychoanalysis and Transversality*. Remarkably, R.A. compares his own speech acts and their difficulties with machinic operations.[24] The intervention of machines in the therapeutic cure allowed for a crucial heterogeneity of relations. This experience allowed Guattari to reconsider the

19 For more on the grid, see Susana Caló, "The Grid," Anthropocene Curriculum, April 23, 2016, https://www.anthropocene-curriculum.org/contribution/the-grid.

20 Guattari, "La Borde," 180.

21 Guattari, "La Borde," 183.

22 See also the brilliant analysis of the patient R.A. in Schmidgen, *Das Unbewusste der Maschinen*, 91–98.

23 Félix Guattari, "Monograph on R.A.," in *Psychoanalysis and Transversality: Texts and Interviews 1955–1971*, trans. Ames Hodges (Los Angeles: Semiotext(e), 2015), 38.

24 "I say 'in short' as if I were a steam machine, both living and dead, that sometimes makes a small sound of steam being released, somewhat regularly, as if it were woken up a little bit by a sound in its dreams of clouds. It does not move but jumps while releasing the words 'in short' in one sentence, then two, then... All of my subsequent 'in shorts' serve the same cause (remember this when you read them)." Félix Guattari, "Collapse of a Life Not Lived. Loss of the 'I': (Extracts from the Journal of R.A.)," in *Psychoanalysis and Transversality: Texts and Interviews 1955–1971*, trans. Ames Hodges (Los Angeles: Semiotext(e), 2015), 42.

25 Guattari and Deleuze, "Balance-Sheet," 91. Emphasis in original.

26 Guattari and Deleuze, "Balance-Sheet," 114.

27 Schmidgen, *Das Unbewusste der Maschinen*, 162 (translation our own).

28 Félix Guattari, "Machinic Heterogenesis," in *Chaosmosis: An Ethico-Aesthetic Paradigm*, trans. Paul Bains and Julian Pefanis (Bloomington: Indiana University Press, 1995), 33.

29 Michael Hardt, "Anti-Oedipus: Part One," Reading Notes on Deleuze and Guattari: *Capitalism & Schizophrenia*, http://people.duke.edu/~hardt/ao1.htm.

structuralist analogy between psyche and machine. For Guattari, machines assumed an equivalent status to the psyche. Such reciprocal interactions went from one machine to another, from psyche to machine. "*Humans are a component part* of the machine, or combine with something else to constitute a machine [...] under specific conditions."[25]

These two cases reveal that Guattari's concept of the machine divorces itself from its conventional confinement to the realm of technics.[26] It instead refers to the productive coupling of heterogeneous components and processes across different domains, such as the social, political, institutional, psychical, and aesthetic. As a theoretical lens, it thus foregrounds the "coaction of human activities, machinic functions, and further processes."[27] This conceptual shift entails a radical revaluation according to which the machine "becomes the prerequisite for technology."[28] Moreover, it puts forth the "properly ontological claim" that "all is machinic."[29]

Cybernetic Ecology

Guattari and Deleuze's elaboration of the machinic thus departs from Lacan's rigid transposition of cybernetics to psychoanalysis by working through more experimental couplings and heterogeneous domains. Yet this entails a proper reckoning with, rather than repudiation of, cybernetics.[30] Indeed, the machine-infused philosophy that the two developed, took up cybernetics' expansion of the machinic, yet with two critical caveats. First, it radicalized the cybernetic principle that feedback opens a machine to its outside, allowing it to incorporate and respond to the contingency of the environment.[31] Accordingly, "the machine no longer faces to the past and determinism, but rather toward rupture, opening, and creativity"[32]—an "intrusion of the radically new" laden with deep political valence via the protests of 1968 France.[33] The concern with "rupture, opening, and creativity" and the recognition of the machinic as a connection of heterogeneous elements also announce the second caveat: Guattari's focus on subjectivity. This rupture on the part of the machinic is itself that which effects subjectivity; as a corollary, machines—including "technological machines of information and communication"—"operate at the heart of human subjectivity, not only within its memory and intelligence, but within its sensibility, affects, and unconscious fantasms."[34]

Thus, although Guattari never engaged with the concept of extimacy, his and Deleuze's "schizoanalytic" machinism does Lacan one better. Traversed not by the tautological alterity of the other but by the heterogeneity of the world, subjectivity becomes truly *extimate*: open to political, social, institutional, aesthetic, as well as technical relations. This extimate production of subjectivity is, in all its valences, machinic.

30 Cf. Andrew Goffey: "Even when [Guattari] is talking about the production of subjectivity in a post-media era, when he is concerned with planetary computerization [...] these ideas nevertheless remain consistent with and are strongly informed by his exploration of the transformations of libidinal energy at play in the kind of psychopathological phenomena—psychosis, schizophrenia, autism, and so on—that were of central importance to him in his lifelong work at the La Borde clinic. The plural semiotic processes which he emphasizes as being operative in the production of subjectivity today [...] are to be understood as indissociable from a growing emphasis in his work on the pathic dimension of subjectivity—a dimension that is emphatically focused on the existential in a manner that simply disappears when people [...] assimilate his work on ideas to do with the machinic to a kind of recondite form of cybernetics." Andrew Goffey, "La Borde and the Analytic Practices of Jean Oury," lecture, Visual Cultures Public Programme, Goldsmiths, University of London, January 30, 2020, London, YouTube, 56:41, https://www.youtube.com/watch?v=iLITiFxAB9E, 13:45–15:25.

31 On cybernetics and the incorporation of feedback and contingency, see Geert Lovink and Yuk Hui, "Cybernetics for the Twenty-First Century: An Interview with Philosopher Yuk Hui," *e-flux Journal*, no. 102 (September 2019), https://www.e-flux.com/journal/102/282271/cybernetics-for-the-twenty-first-century-an-interview-with-philosopher-yuk-hui; expounded at length in Yuk Hui, *Recursivity and Contingency*, Media Philosophy (London: Rowman & Littlefield, 2019).

32 Schmidgen, *Das Unbewusste der Maschinen*, 163 (translation our own).

33 Schmidgen, *Das Unbewusste der Maschinen*, 65 (translation our own); see also 58–61. For the political roots informing institutional psychotherapy more generally, see Camille

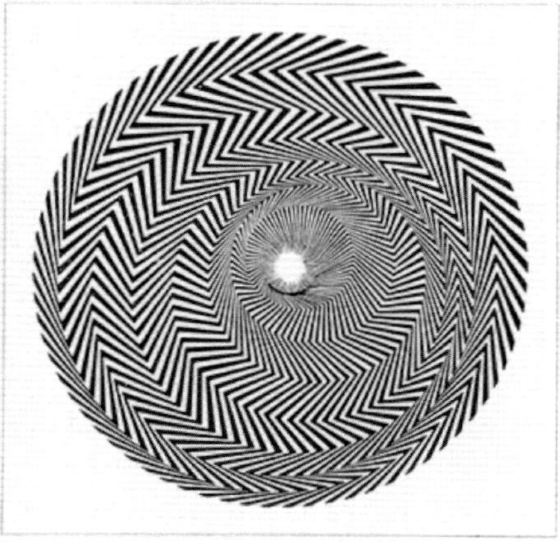

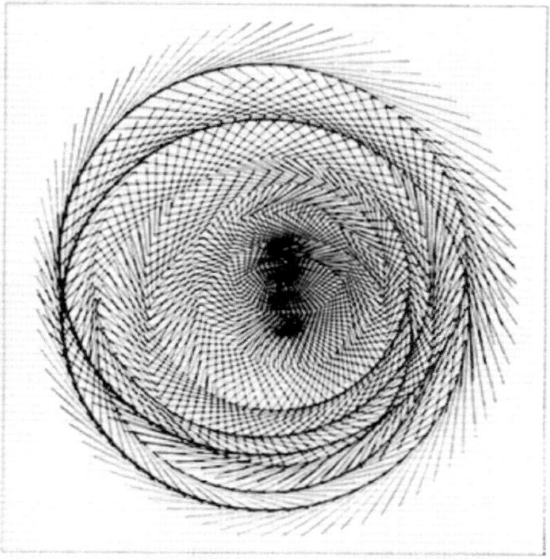

Bridget Riley, *Blaze I*, 1962. Emulsion on board. Leigh Hendricks, *Variations on a Theme from Bridget Riley*, 1968. Computer graphic. In: Jasia Reichardt, ed., *Cybernetic Serendipity: The Computer and the Arts* (New York: Studio International, 1969)

In this light, the digitization of the lifeworld and the concomitant extension of market forms of organization merit special attention.[35] The Deleuzo-Guattarian machinist lens affords a perspective that apprehends the psychical, technical, economic, etc. as constituting one immanent field. As institutional psychotherapy and Gestalt psychology before it recognized, behavior emerges from the constitutive reciprocities between the poles conventionally understood as organism and environment.

Today, companionship in digital spaces has become a fact of everyday life. The imbrication of information technology in the most minute gestures, affects, and habits reveals a spectrum of *machinic ethological hieroglyphs:* modes of behavior and attitudes that testify to a psychical, digital, and social infrastructure. Their symptomatic expression through articulations of *intimacy* proves their incommensurability with the available psycholinguistic inventory. For one thing, the inherence of digital mediation in interpersonal intimacy has taken on the character of a sine qua non relation accepted and felt at the level of the cultural unconscious. Deprivation—"without information and limited to the people around"—induces depressive isolation.[36] Furthermore, intimacy prevails as a conceptual modality for articulating affects of the emergent socio-technical milieu. It recurs, for instance, in nearly half of the entries in New Feelings, a column in the Snapchat-sponsored magazine *Real Life*, "devoted to the desires,

Robcis, "François Tosquelles and the Psychiatric Revolution in Postwar France," *Constellations* 23, no. 2 (June 2016): 212–22.

34 Félix Guattari, "On the Production of Subjectivity," in *Chaosmosis: An Ethico-Aesthetic Paradigm* (Bloomington: Indiana University Press, 1995), 4.

35 A notable contribution on the concomitance of market forms of organization and information technology is Claudio U. Ciborra, "Reframing the Role of Computers in Organizations: The Transaction Costs Approach," in *Proceedings of the 6th International Conference on Information Systems (ICIS), Indianapolis, Indiana, 1985* (Atlanta: Association for Information Systems, 1985), 57–69, http://aisel.aisnet.org/icis1985/9.

36 "Isolation," the world UNPLUGGED, 2019, last modified March 3, 2011, https://theworldunplugged.wordpress.com/emotion/isolation.

37 "New Feelings," *Real Life*, https://reallifemag.com/tag/new-feelings.

38 Theodore Roszak, *From Satori to Silicon Valley* (Stanford: Stanford University, 2000), electronic edition, Making the Macintosh: Technology and Culture in Silicon Valley, https://web.stanford.edu/dept/SUL/sites/mac/primary/docs/satori/index.html.

39 Fred Turner, *From Counterculture to Cyberculture: Stewart Brand, the Whole Earth Network, and the Rise of Digital Utopianism* (Chicago: University of Chicago Press, 2006), 38.

40 Stewart Brand et al., eds., "Purpose," *Whole Earth Catalog*, 1968–72.

41 Turner, *From Counterculture to Cyberculture*, 31; see also Theodore Roszak, "The Short Cut to Satori" in Roszak, *From Satori to Silicon* Valley, https://web.stanford.edu/dept/SUL/sites/mac/primary/docs/satori/short.html.

42 Richard Brautigan, "All Watched Over by Machines of Loving Grace," in *All Watched Over by Machines of Loving Grace* (San

moods, pathologies, and identifications that rarely had names before digital media."[37] These entries name the new configurations of intimacy, which no longer exist within the continuity of inner- or intersubjective space, as well as new symptoms of a subjectivity conditioned by new forms of machinic entanglements.

A tension emerges from this situation. On one hand, an ongoing, thorough recompiling of the psychosocial stack with evermore technological components immanentizes it with the digital. It thereby undermines traditional models of interiority. On the other hand, the subjective experience of this process apparently refers back to intimacy—yet what kind of intimacy is at stake here? There is a linguistic and experiential persistence of intimacy as longing for interiority even as its very conditions are in considerable flux. In an apparent paradox, in the midst of a reconfiguration revealing itself in all its extimate and machinic dimensions, classical notions of interiority—and with it, intimacy—resurface.

This coexistence dates back to the origins of the current technological regime. The "technophiliac"[38] "new communalists" of the American 1960s counterculture sought to effect a transition away from the 1950s Fordist disciplinary *dispositif*, its bureaucracy, technocracy, and proffered model of a "drab, psychologically distressing adulthood,"[39] toward a "realm of intimate, personal power."[40] Having eschewed the confrontational politics of social change, they "turned inward, toward questions of consciousness and interpersonal intimacy, and toward small-scale tools such as LSD or rock music as ways to enhance both."[41] Evidently, the critique of the prevailing technocracy had less to do with technology as such and more to do with the perceived unfeeling rigidity of the homogenizing apparatus. Against the latter, cybernetics offered an epistemological framework that rendered technology more supple, appropriately scaled, and responsive to the social, human, and natural. This liquefaction applied to the perception, use, and development of technology. Communes, acid trips, festivals: these sites prototyped, demoed, and beta-tested the "cybernetic ecology"[42] that would recuperate their failure.[43] The personal, networked computer incorporated and reproduced aspects of this milieu.

The relevance of this genealogy lies in the fact that the technophiles of the Californian counterculture not only occupied a remarkably influential niche but also essentially "endeavored [...] to produce a new type of subjectivity" as a sociopolitical strategy.[44] Potent technical machines such as personal computing, psychedelics, as well as multimedia and improvisational arts infused an experimental milieu. These elements were assembled into machines of subjectivation. Conceptually, the psychic infrastructure underlying such a cultural, political, and ethical utopia fundamentally implies the extimacy of processes of subjectivation.

Francisco: Communication Company, 1967), http://www.brautigan.net/machines.html#28.

43 See Turner, *From Counterculture to Cyberculture*, 146–48; this was also a retroactive move, as for example in Stewart Brand, "We Owe It All To The Hippies," *TIME*, Spring 1995, http://members.aye.net/~hippie/hippie/special_.htm.

44 Guattari, "La Borde," 180.

Thus, this section of the counterculture evinces some striking commonalities in terms of problematizations and strategies with Guattarian theory and practice. Yet the dispositif that crystallized from the assemblage gestating west of the Atlantic would come to mobilize its technical resources in order to reinscribe models of interiority. Two factors would play decisive roles in this occurrence.

One, the personal computing "revolution," like the counterculture before it, drew on American cultural reservoirs of individualism and frontier mythology and also naturalized mid-twentieth-century capitalist relations. Economically and discursively, this heritage would prove fruitful for the incipient digital intimacy by offering a solution to—that is, deferring—a problem that cybernetics and its technologies had continuously contended with: the potency of the cybernetic "universal solvent."[45] Applied to social and technical machineries, it rendered them more flexible, supple, and responsive, but that same agnosticism to the boundaries between entities, whether cultural or natural, also extended to and put individuals' bodily or subjective "boundaries [. . .] up for grabs," thus seriously imperiling the classical model of liberal subjectivity.[46] Incorporating an individualism steeped in a mythological reservoir fortified the construction of psychological unity, the illusion "that mask[s] the cybernetic reality" of interconnected machines at once constituting and threatening to erode the subject.[47] Messages on the WELL, an early internet

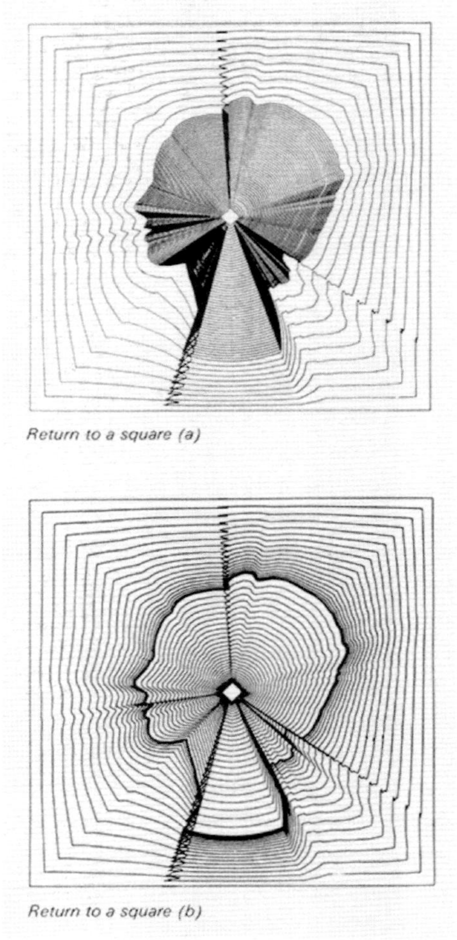

Masao Komura and Kunio Yamanaka, *Return to Square*, 1968. Computer graphic. In: Jasia Reichardt, ed., *Cybernetic Serendipity: The Computer and the Arts* (New York: Studio International, 1969)

45 N. Katherine Hayles, *How We Became Posthuman: Virtual Bodies in Cybernetics, Literature, and Informatics* (Chicago: University of Chicago Press, 1999), 85.

46 Hayles, *How We Became Posthuman*, 85.

47 Hayles, *How We Became Posthuman*, 109.

48 Turner, *From Counterculture to Cyberculture*, 158–59; see also 146–47.

49 Philip E. Agre, "Beyond the Mirror World," in *Technology and Privacy: The New Landscape*, ed. Philip E. Agre and Marc Rotenberg (Cambridge, MA: MIT Press, 1997), 31, modified version available at http://polaris.gseis.ucla.edu/pagre/mirror.html; see also e.g. Charles Irby et al., *A Methodology for User Interface Design* (Palo Alto: Xerox Information Technology Group, Systems Development Division, 1977), http://www.bitsavers.org/pdf/xerox/sdd/A_Methodology_for_User_Interface_Design_Jan77.pdf.

50 Gilles Deleuze, "What Is a Dispositif?," in *Michel Foucault, Philosopher: Essays*, ed. and trans. Timothy J. Armstrong (New York: Harvester Wheatsheaf, 1992), 159–68.

51 See for instance Roderick Perkins, Dan Smith Keller, and Frank Ludolph, "Inventing the Lisa User Interface," *Interactions* 4, no. 1 (January 1997): 40–53; see also David Canfield Smith et al., "Designing the Star User Interface," *BYTE*, April 1982, 242–82, https://tech-insider.org/star/research/acrobat/8204.pdf.

52 See Philip E. Agre, "Surveillance and Capture: Two Models of Privacy," *Information Society* 10, no. 2 (April 1994): 101–27; see also Bernhard Rieder, "Beyond Surveillance: How Do Markets and Algorithms 'Think'?," *Le Foucaldien* 3, no. 1 (2017), https://doi.org/10.16995/lefou.30.

53 On the PC dispositif in general and its foregrounding of user agency, see Udi ben Arie and Noam Knoeller, "The Holodeck Is All Around Us—Interface Dispositifs in Interactive Digital Storytelling,"

discussion board, eagerly assimilated the networked infrastructure of the milieu to content reminiscing on the days in the commune, combining both in an "experience of interpersonal intimacy."[48] A kind of—partially retroactive—cultural-technological causality aligned medium and message.

Two, personal computing embedded in the extant computing industry, in the logistics of its material development, in the knowledge formations and practices it imported, and in the market pressures its research and development teams responded to. Computing had long been "assimilated" and "essentialized" to bureaucratic procedures and a quasi-Taylorist information processing paradigm.[49] Thus, the development of the graphical user interface—the vanguard apparatus by which personal computing would modify "lines of sight" and produce the new modes of perception necessary for the sought-after resubjectivation process[50]—modeled the user as an individual, specifically a bureaucratic office worker.[51] Concretely, work-rationalization and step diagrams of office work laid the procedural and metaphoric groundwork for the human-computer interaction paradigm. Critically, this alignment of market logics and individual subjectivation by way of digital interfaces proceeds by the disambiguation of potential actions and concepts, thus externalizing them from the user.[52]

This counterculture–Silicon Valley overlap thus quite effectively postponed the problem posed by the cybernetic dissolution of subjective in- and exteriority. The most comprehensive manifestation of this "Californian solution" was the desktop "PC interface dispositif" that served as the dominant human-computer interaction paradigm from its market debut in the 1980s until the early twenty-first century.[53] This dispositif of interface design places user agency in the foreground and produces a spatiality consistently external to the user.

The second (and ongoing) wave of personal computing is based on smaller, even more supple devices such as smartphones and new business models of platformization. Premised on the datafication and capture of life and affective experience via gestural interfaces as well as near-ubiquitous and contextually aware computing, the second wave reorders the infrastructure constituting the user's in- and exteriority.[54] In its interface design logics, it holds on to Californian technological libertarianism's insistence on the user as an individual, thus enabling the feeling and divulgence of intimacy. At the same time, it exploits the cybernetic dissolution of said individual via the infrastructure of algorithmicized extimacy it installs.

Digital Ethology

In recent years, interface design has gravitated toward minimalism, receding into the background in order to prioritize the presentation of content. Seeking attention, interfaces also employ a variety

in *Interactive Digital Narrative: History, Theory and Practice*, ed. Hartmut Koenitz et al., Routledge Studies in European Communication Research and Education 7 (New York: Routledge, 2015), 51–66. Regarding consistency as a design principle, see e.g. Apple Computer, Inc., *Macintosh Human Interface Guidelines*, Apple Technical Library (Reading, MA: Addison-Wesley Publishing Company, 1992); on the point of consistency and the interface's relation to quasi-Taylorism, see Don Gentner and Jakob Nielsen, "The Anti-Mac User Interface," *Communications of the ACM* 39, no. 8, August 1996, 70–82, Nielsen Norman Group, https://www.nngroup.com/articles/anti-mac-interface.

54 Soenke Zehle, "The Autonomy of Gesture: Of Lifestream Logistics and Playful Profanations," *Distinktion: Journal of Social Theory* 13, no. 3 (December 2012): 340–53.

of hooking mechanisms such as infinite streams and auto-playing videos.[55] These changes effectively recant the exteriority characteristic of the PC interface vis-à-vis the user. The immediacy of gestural interaction and pervasive nudging habituates users to the continuous and continually generated stream of content, curated to enable evermore-capable tracking mechanisms. Where once interfaces leaned heavily on skeuomorphic cues to afford specific interactions, now, nudges, gamification, and habituation take over as compulsion mechanisms of a behaviorist paradigm.[56] One scrolls "adrift among the consciousness of others" on Twitter's platformized disorientation of "intimacy turned malignant" or floats through an endless river of infinitely looping TikTok videos.[57] Except *one* is not doing these things: it is not a subject interacting with objects, whether a person using their phone, an organism with its environment. Rather, the machines in question consist of database logic and reflexes, vision and haptics, institutions and dopamine release: moments of intensifying feedback loops that render any attempt to distinguish between a psychic interiority of the user and an exterior space of the infrastructure a blurry mess.

This "machine zone," wherein "affect and behavior at once condition and are conditioned by the system," employs the same principles that Natasha Dow Schüll finds at work in machine gambling.[58] The "desubjectifying effects" of this sort of frictionless "uninterrupted flow" interface design draw on an experience of near-perfect contingency.[59] Yet whereas machine gambling aims this at "self-erasure,"[60] platform infrastructure maintains a thin membrane around a space of interiority. Elements that require discrete decisions, from profiles to post fields and reaction buttons, constitute the links of this membrane. Its selective permeability conditions the circulation of intimate thoughts, experiences, and inclinations—which, in turn, filter into other users' streams. This interfacial logic performs a critical mediating and productive role in effecting new machines that operate across psychical and social domains.

Whether the Instagram feed showing friends' family homes[61] or Tinder's extension of the "commercial revolution" of "location-based services" to "the discovery of people,"[62] the organizing principle corresponds to what Schüll refers to as "technologically contrived contingency management."[63] Here, the disambiguation effected by software plays a critical role. Schüll compares machine gambling and its straightforward stakes and concepts to the more uncertain, experiential, and messy labor of emotion or risk management that societal transformations thrust upon unprepared subjects.[64] The economic calculus of the reduced transaction costs offered by an informatized market for the "discovery of people"—or, for that matter, topics or events—enters into this social field. This same social field is undergoing liquefaction

55 For the paradigmatic text of the "hooking" approach to software design see Nir Eyal, *Hooked: How to Build Habit-Forming Products* (New York: Portfolio, 2014); for an overview, see also Chauncey Neyman, "A Survey of Addictive Software Design" (bachelor's thesis, California Polytechnic State University, 2017), https://digitalcommons.calpoly.edu/cscsp/111.

56 On the shift from skeuomorphic to "flat" design generally, see e.g. Ivan Burmistrov et al., "Flat Design vs Traditional Design: Comparative Experimental Study," in *Human-Computer Interaction – INTERACT 2015: 15th IFIP TC 13 International Conference, Bamberg, Germany, September 14–18, 2015, Proceedings*, ed. Julio Abascal et al., vol. 2, Lecture Notes in Computer Science 9297 (Cham: Springer International, 2015), 106–14. Note that as of 2020, this aesthetic trend is less extreme than at the initial shift to "flatness." For a critical recounting of broad shifts in interface design, see also Cade Diehm, "On Weaponised Design," Tactical Tech, 2018, https://ourdataourselves.tacticaltech.org/posts/30-on-weaponised-design.

57 Chris Randle, "New Feelings: Roach Complex," *Real Life*, June 13, 2019, https://reallifemag.com/new-feelings-roach-complex.

58 Natasha Dow Schüll, *Addiction by Design: Machine Gambling in Las Vegas* (Princeton: Princeton University Press, 2012), 165. This connection is made explicitly in Shoshana Zuboff, *The Age of Surveillance Capitalism: The Fight for a Human Future at the New Frontier of Power* (New York: PublicAffairs, 2019).

59 Schüll, *Addiction by Design*, 166, 171.

60 Schüll, *Addiction by Design*, 171.

61 Alexandra Molotkow, "New Feelings: Selfish Intimacy," *Real Life*, January 17, 2019, https://reallifemag.com/new-feelings-selfish-intimacy.

62 Gareth Tyson et al., "A First Look at User Activity

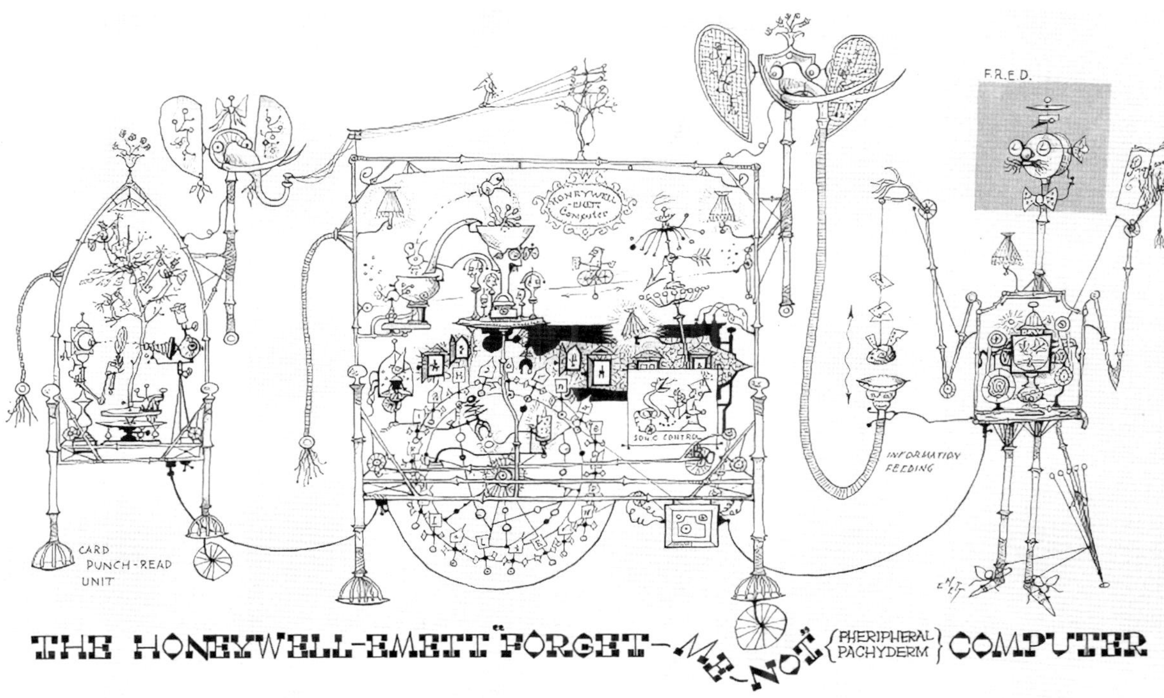

Frederick Rowland Emett, *The Honeywell-Emett "Forget-Me-Not" (Pheripheral Pachyderm) Computer*, 1966. Pen ink and watercolor with bodycolor

and marketization via the entangled vectors of progressive-ization, neoliberalism, and digitization. The ordering of disambiguating grammars and personalized matching algorithms interface to this complexity; gamification and habituation design patterns not only acclimate users to this socio-technical milieu, but they also manage the anxiety and uncertainty of intersubjective contingency.[65] For instance, as the erosion of heteropatriarchal social scripts grants subjective freedom and bodily safety, it leaves a psychosocial vacuum of uncertainty effectively responsible-ized to the individual. Platforms, with their algorithmic management of contingency, readily take up the task of reconfiguring flows of desire by disintermediating the components of the classic social milieu.[66] This cybernetic

on Tinder," in *Proceedings of the 2016 IEEE/ACM International Conference on Advances in Social Networks Analysis and Mining: ASONAM 2016*, ed. Ravi Kumar, James Caverlee, and Hanghang Tong (Piscataway, NJ: Institute of Electrical and Electronics Engineers), 461.

63 Schüll, *Addiction by Design*, 192.

64 Natasha Dow Schüll, "Gambled Away: Liquidating Life," chap. 7 in Schüll, *Addiction by Design*, 189–209, especially 194–95.

65 On the question of interfaciality and complexity, see Patricia Reed, "Uncertainty, Hypothesis, Interface," in "Scientific Romance," ed. Beatriz Ortega Botas, special issue, *_AH*, no. 00 (March 2017), archived March 22, 2018, https://web.archive.org/web/20180322080411/http://www.ah-journal.net/issues/00/uncertainty-hypothesis-interface; Patricia Reed, "Orientation in a Big World: On the Necessity of Horizonless Perspectives," *e-flux Journal*, no. 101 (June 2019), https://www.e-flux.com/journal/101/273343/orientation-in-a-big-world-on-the-necessity-of-horizonless-perspectives; and Benjamin H. Bratton, "Interface Layer," in *The Stack: On Software and Sovereignty* (Cambridge, MA: MIT Press, 2016), 219–49.

66 Kate Julian, "Why Are Young People Having So Little Sex?," *Atlantic*, December 2018, https://www.theatlantic.com/magazine/archive/2018/12/the-sex-recession/573949; Michael J. Rosenfeld, Reuben J. Thomas, and Sonia Hausen, "Disintermediating Your Friends: How Online Dating in the United States Displaces Other Ways of Meeting," *Proceedings of the National Academy of Sciences* 116, no. 36 (September 3, 2019): 17753, https://doi.org/10.1073/pnas.1908630116. On the subject of love, algorithms, and risk, see also Anna Longo, "Love in the Age of Algorithms," trans. Gila Walker, *Glass Bead*, no. 2 (2019), https://www.glass-bead.org/article/love-in-the-age-of-algorithms.

solvent technicizes, dissolves, and reassembles the social milieu and individual alike.

The management of contingency thus concerns the exposure to and conditions of experienced intimacy. The digital infrastructure, by virtue of its increasing capacity to incorporate and learn from the contingency of the psychosocial *Umwelt*—processes by which it also comes to embody and produce contingency itself—begins to take on a wisp of the machine in its Guattarian sense.[67] This widespread and privileged[68] access to the production of intimacy approves a diagnostic in terms of Deleuze and Guattari's thesis that a "social attitude vis-à-vis machines [. . .] expresses [. . .] the position of desire in the infrastructure."[69]

In this light, the cultivation of spaces of interiority appears significant. This adds causal linkage to a concurrence Guattari already noted in the early 1990s, of the "massive development of machinic productions of subjectivity" with its "conservative reterritorialisations."[70] The nostalgic warmth and familiar spatiality of intimacy readily offer themselves for cliché experience and commodification. Further, the careful calibration of this space as one of continuous feedback suggests that "technologically contrived intimacy management" may be *the desired projection of machinic extimacy* today.[71]

The scrolling machine is in motion and propels the stream. Some of the endless banalities, exhortations, and absurdities affect you. They do so because they connect to other machines: some long dormant, others sputtering on and off again in periods of perhaps days or weeks, and some constantly running background processes. Each of these connects to vast reticulations that traverse affects, social constellations, markets, information logics, and memories. Because you cannot perceive the breadth of this space, your experience forms at a fold onto which its vast multidimensionality is projected. The projection apparatus is another machine (as everywhere) with two unstable poles, the intimate and the extimate.

Here, Lacan's point that extimacy is not a mere counterpart to intimacy takes on renewed significance. Psychoanalysis, as revisited by Guattari, thus provides a fertile ground to think about extimacy as an artificially produced space, open in all directions, and traversed by machines. The logic of this space proves that when machines shape our behavior, behavioral patterns have already constituted amalgamations of machines. The desiring machines of intimacy operate in parallel and in countermotion to extimacy machines. Trying to question the effective topology of the psyche modulated, maintained, and formed by digital operations means, today maybe more than ever, addressing the section of transversal space folded in and through psychic machines.

67 For further treatment of digital technology through a Guattarian lens, see Andrew Goffey, "Micropolitics of Software: Rethinking Programming Practices and the Production of Subjectivity," lecture, Subjectivity, Art and Data, Royal Holloway, University of London, 2018, London, YouTube, 1:42:57, https://www.youtube.com/watch?v=9bqxsmFo72k; see also e.g. Simon Crowe, "Micropolitics of a Recommender System – Machine Learning and the Machinic Unconscious," *Spheres*, no. 5 (November 2019), http://spheres-journal.org/micropolitics-of-a-recommender-system-machine-learning-and-the-machinic-unconscious.

68 For an example of the privileged access that the digital infrastructure holds to the "intimacy" of the individual, see Suzannah Showler, "New Feelings: Screen Protectiveness," *Real Life*, February 19, 2019, https://reallifemag.com/new-feelings-screen-protectiveness.

69 Guattari and Deleuze, "Balance-Sheet," 114.

70 Guattari, "On the Production of Subjectivity," 2–3.

71 The phrase "machinic extimacy" has also been used in a Lacanian study by Marc von Boemken dismantling apocalyptic fantasies vis-à-vis technology: Marc von Boemken, *Machinic Extimacy: A Critical Mediation on the Dialectics of Technological Apocalypse* (Saarbrücken: VDM Verlag, 2009).

This document was leaked on 08.06.2021 by an informant who will remain anonymous for her protection. It concerns NASA's first internal report regarding the infamous "Event PHV554"—issued under its highest restriction code, CTSA—containing what it declares to be a copy of the alien source material in full.

Internal Report: Event PHV554 06.06.2021
Cockburn, Western Australia

We're still investigating how it got here.

The site reveals a crater of 3,900ft/1,200m in diameter, from which we calculate that the object's speed upon impact was 30,000mph/12.8km/s. The shock wave it generated covered over 200mi²/518km², causing widespread devastation to the area. Death and injury counts are high and rising, since no evacuation warning was issued due to the absence of prior observations from our satellites.

Under normal circumstances the absence of prior observations would be highly anomalous, as all Earth-threatening meteorites are currently detected decades ahead of entering Earth's atmosphere.

In this case, however, this is explicated by the itself inexplicable fact that the crater is of course inverted, measuring 560ft/170m in height (rather than in depth), with a rim of 148ft/45m below (rather than above) the surrounding area.

Meaning the meteorite came from within the Earth, rather than from outer space.

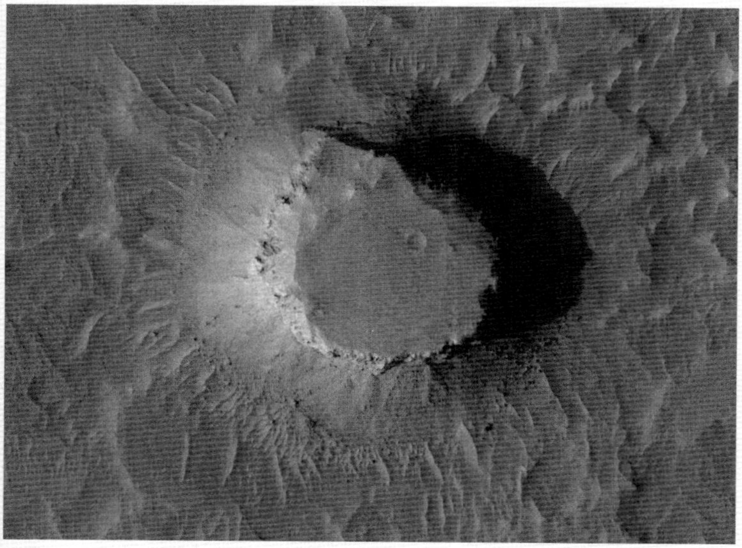

Grayscale image of Site T Source: MSI on Sentinel-2B
Resolution: 20m (Downsampled)

OP SECRET ATOMAL (CTSA) COSMIC TOP SECRET ATOMAL (CTSA) COSMIC TOP SECRET ATOMAL (CTSA) COSMIC TOP SECRET ATOMAL (CTSA) COSMIC TOP SECRE

We've simulated all possible scenarios based on our geological models, which confirm our initial hypothesis with a confidence coefficient of .99:

Under no real-world circumstances is it possible for an object coming *from within the Earth* to reach the speed required for it to create a crater of this magnitude.

In conclusion, we have within the current state of knowledge no explanation for Event PHV554.

Popular opinion has conjectured some sort of quantum portal, though this theory is entirely ungrounded as of now.

The object we found at the center of the crater consists of an oval-shaped outer shell made of an unidentified semi-solid/viscous substance, containing within it –crumpled up in a ball– a six-page document with "interpretive notes" concerning a so-called "LUNAR LIBRARY II" (LL2). The author(s) claim to have the LL2 in their possession.

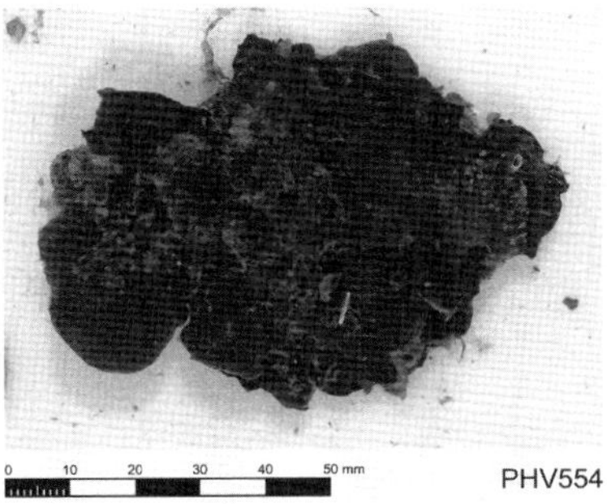

PHV554

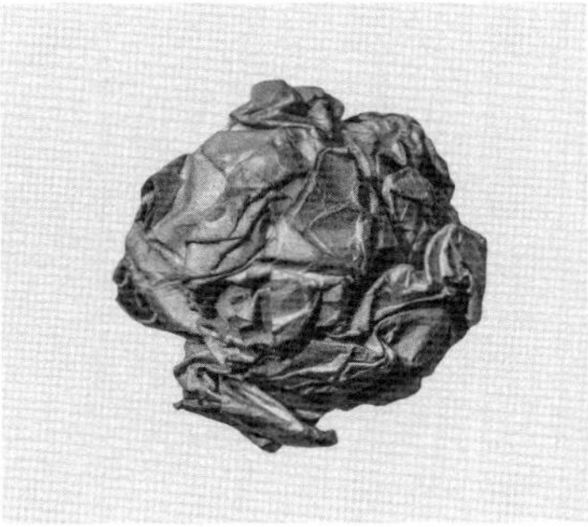

According to our intelligence, there exists only one LUNAR LIBRARY™: a series of archives from the Arch Mission Foundation designed to preserve the records of our civilization for up to billions of years. It was installed in the SpaceIL "Beresheet" lunar lander, which crashed on the Moon in April 2019.

Arch Mission plans to launch a second instalment of the LUNAR LIBRARY™ in 2022.

In the document we retrieved from the crater, the LL2 is indeed dated "2022", i.e. one year from now, which would suggest some form of time travel and the author(s)' possession of an archive that does not yet exist.

While we do not exclude the possibility that the document is a hoax, this would not yet explain the geological/cosmic event that created the inverted crater at Site T.

The heading of the document states "draft" and its rather fragmentary, at times nonsensical nature gives the impression it is unfinished.

Attached is a transcribed and unedited copy of the document in full. We've annotated it in the margins, mostly for clarification insofar as we're able to provide any at this stage.

Attachment 1

LUNAR LIBRARY II: interpretive notes (draft)

We're pretty clear on how it got here.

All signs point to a smooth interworld passage through The Cakewalk: a relatively wide, slow-progression portal with a propagation capacity of about 1.38, most of which was consumed.

Commented PHV554 Report_Admin:
This may refer to:

"Cakewalk": a computer software for recording and making music.

"The Cakewalk": a dance that was developed by plantation slaves in the Southern U.S. in the mid-19th century, which is understood to have enabled slaves to covertly mock their owners without getting punished. James Baldwin states that it was meant "to satirize the competing culture of supposedly 'superior' whites, who were able to dismiss its threat in their own minds by considering it as a simple performance which existed for their own pleasure".

The documentation is spectacular.
Although what we can glean from it is blowing our minds:

A highly atomized /sub/species-enclosed yet fully operational anamorphic subjecthood, with degeneration catalyzing at 1492 (homo-unidirectional-static [HUS] calendar), that blew at HUS2020–21 with the (late) arrival of auto-induced biotechnology "Corona Pandemic" (not to be mistaken for infectious agent SARS-CoV-2: the virus that ran through it).

As far as we can tell, this was the first biotechnology to become toxic/lethal to its makers, and it's weirding us out.

♥♥♥♥♥♥♥♥♥♥♥♥♥♥♥♥♥♥♥♥♥♥♥♥♥♥♥♥♥♥♥♥♥

LL2 source notes

The makers generally describe the LL2 as *both*:
- "a 30-million-page library aimed at preserving human civilization"
AND
- "a backup to planet Earth"

This was long thought to be a joke given the obvious incongruity of asserting any correspondence between the two, but recent revisionary theories suggest it's more likely attributable to the self-synecdochal tendencies of the authors. The latter would be in tune with their pathological profile, in which this pattern is pervasive (common instances include "white heterosexual males" for "humanity", "0.00000001% of western culture" for "culture", and "profit-generating pharmaceutical industry" for "health care", for instance).

The LL2 is supposed to contain "scientific, cultural, and historical information", though we have yet to understand the partitional logic here.

The LL2 rolled in through portal branch (͡°(͡° ͜ʖ(͡° ͜ʖ ͡°)ʖ ͡°) ͡°), which we haven't been shy to admit we're kind of sweet on, since it transports the largest load of Anthropo-not-seen persons, whose high sensual (dis)charge may have been what caused LL2 to literally "slip through", notwithstanding its (f)rigidity.

Commented PHV554 Report_Admin:
This term was coined by Peruvian anthropologist Marisol de la Cadena. She defines it as "the world-making process by which heterogeneous worlds that did not make themselves through the division between humans and nonhumans—nor necessarily conceived as such the different entities in their assemblages—were both obliged into that distinction and exceeded it. The anthropo-not-seen was thus the process of destruction of these worlds and the impossibility of such destruction". https://www.haujournal.org/index.php/hau/article/view/hau4.2.013

♥♥♥♥♥♥♥♥♥♥♥♥♥♥♥♥♥♥♥♥♥♥♥♥♥♥♥♥♥♥♥♥♥

What follows are some preliminary notes based on our reading of the LL2's contents, specifically concerning the roughly 500 HUS-years in which the biotechnology "Corona Pandemic" was manufactured, and how this process led to the by now well-known cosmic events in which it debouched.

♥♥♥♥♥♥♥♥♥♥♥♥♥♥♥♥♥♥♥♥♥♥♥♥♥♥♥♥♥♥♥♥♥♥

The manufacturing process was extremely elaborate, beginning in HUS1492, and catapulting quite quickly from there. Not coincidentally, HUS1492 is exactly when worlds started to end, according to our records.

A symbiotic assemblage comprising human and fungal voices of the circumboreal Matsutake-complex explains how the golden formula for smooth expansion in the next 500 HUS-years was found through the development of the European colonial plantation. As they say, it basically came down to this: *exterminate local people and plants;* prepare now empty, unclaimed land; and bring in exotic and isolated labor and crops for production.*

*The exceedingly high frequency with which the word "people" in the LL2 refers exclusively to humans (especially in those timespace areas where the atomization of the /sub/species-enclosed anamorphic subjecthood peaks) leads us to conclude that personhood was afforded primarily to humans; hence the awkward use of "and" as a conjunctive between "people" and "plants".

Popular for its simplicity—which made possible its implementation at limitlessly large scales—this model became an inspiration for later industrialization and modernization. Indeed, we find evidence of it all over the place, spreading exponentially and in numerous variations,° quite like a cancer.

°In the LL2, we find that the congeneric paradigms of conquest, enclosure, and private property became the model for many things, not just industry —including for instance what came to pass for "love", as we read in the many instances describing feminized bodies as metaphors for land, which must be possessed, enclosed, gated, and defended, for others from elsewhere might come in to steal it.

Obviously SARS-CoV-2 would have been quite harmless were it not for the debilitated and debilitating infrastructures that proliferated in this process, which are what manufactured its *becoming-pandemic* (first by the reduction of biodiversity as a result of deforestation, mining, industrial agriculture etc., all of which work wonders to catalyze the spread of pathogens; and second by its fast circulation thanks to global extraction/production/distribution channels, as well as mass urbanization and the extortionate growth of global air travel), altogether creating a perfect storm for the spillover of diseases.=

=In fact, the reason why we describe the arrival of Corona Pandemic as "late" is because, for decades, scientists had been issuing warnings about the danger of pandemic outbreaks (which they predicted would occur earlier), precisely given the abovementioned combination of environmental collapse and industrial globalization. They also warned that preservation of the status quo would likely lead to more and possibly worse pandemics.

In excess of this, it's clear that part and parcel of the unceasing Genocide was its co-constitutive Epistemicide, such that gathering seeds, knowledge, and stories—and passing them along across multitudes—gave way to the linear, progressive, Time's-(killing)-arrow mode of the Techno-Heroic, of which the witch hunts that went on from HUS-15th C. until the end of documentation at HUS2022 are a good example.

This genealogical detachment reduced deep-time epistemic interfacing to such a degree that, increasingly, human populations started coming in like newborn adults: all the intelligence there, but no knowledge; all the perceptual capacity, but no sense.

And not only on an epistemic, but also on a geophysiological level we see a remarkable drive to self-confinement—long before the 2020–21 pandemic enforced it by law. Starting with the Scientific Involution that took place early in the 500 HUS-years in question, we see thought collectives develop trained incapacities such as the mechanistic worldview, which hallucinated planet Earth (and the entire universe) as a dead, controllable object: a machine. This hallucination appears to have been quite powerful, and while largely willed (if with differing degrees of awareness), it induced a lethal ontological disconnect that compelled those it

Commented PHV554 Report_Admin: We've found two of the phrases below (which the authors have neglected to place in quotation marks or properly credit, as is the case throughout the entire document) on p.39 of a book by American anthropologist Anna Lowenhaupt Tsing, titled "The Mushroom at the End of the World" (2015).

Commented PHV554 Report_Admin: Possibly paraphrased from an article published in July 2020 by the fictional artist Claire Fontaine, titled "THE VIRUS IS OUR IDEA OF OURSELVES".

Commented PHV554 Report_Admin: We've found very similar wording in an article published in April 2020 by the Intergovernmental Science-Policy Platform on Biodiversity and Ecosystem Services (IPBES), titled "COVID-19 Stimulus Measures Must Save Lives, Protect Livelihoods, and Safeguard Nature to Reduce the Risk of Future Pandemics".

Commented PHV554 Report_Admin: This term was coined by Portuguese sociologist Boaventura de Sousa Santos, who defines it as "the destruction of the knowledge and cultures of [indigenous] populations, of their memories and ancestral links and their manner of relating to others and to nature". (p.18 in "Epistemologies of the South and the Future", 2016).

Commented PHV554 Report_Admin: Direct quote from a 1986 essay by Ursula Le Guin, titled "The Carrier Bag Theory of Fiction".

Commented PHV554 Report_Admin: This phrase appears in the science-fiction novel "Midnight Robber" (2000) by the Jamaican-born Canadian writer Nalo Hopkinson.

Commented PHV554 Report_Admin: Originating from the German "Denkkollektiv", this term was coined by Polish-Jewish microbiologist Ludwik Fleck, who uses it in his 1935 book "Genesis and Development of a Scientific Fact" to claim that cognition is a collective activity.

Commented PHV554 Report_Admin: We've found this phrase in American poet and literary theorist Kenneth Burke's book "Permanence and Change", where he defines it as "a state of affairs whereby one's very abilities can function as blindnesses", such that "a way of seeing is also a way of not seeing".

consumed to embark on a most absurd endeavor: the waging of a centuries-long war *against* "nature".§

§In the hallucination, "nature" didn't include those who were hallucinating or the objects they considered to be theirs according to their notion of species-bound authorship.

§§Those most deeply affected by the war and those whose livability was destroyed by it were not those who were waging it, at least not within the narrowest of windows we can think personal pronouns into.

This combative and competitive impulse for survival of the bounded self as distinct from (that of) the living environment led to a further decrease in interaction among interdependent communities, thereby minimizing the differences that diversify the materials and processes available for sharing: the principal condition for life. Diminished promiscuity eventually became straight-out incestual, with the /sub/species-enclosed subjecthood becoming very ill and foisting upon itself increasingly specialized habitats and niche conditions for its subsistence. That these characteristics are the strongest indicators of a species' vulnerability to extinction was known: the LL2 contains numerous publications by a loud yet (strangely) unheard holobiont (comprising eukaryotic as well as prokaryotic lifeforms), who enunciated its onto-logic in a series of documents signed "Lynn Margulis". In crystal-clear language, they make plain how and why the more specialized a species' habitat, the more impoverished its (encoded and learned) adaptation capabilities become, which almost always leaves it defenseless as soon as the environment changes. Which of course it did.#

Commented PHV554 Report_Admin: American evolutionary theorist known for her highly contested claim that symbiosis (rather than competition) is the central force behind the evolution of new species. Her views run counter to those of neo-Darwinian biologists, whose gene-centered theory of evolution as "survival of the fittest" posits that living agents evolve by competing for exclusive advantage over the limited resources in their environments.

#It's also true that this particular /sub/species had more than one strong indicator of vulnerability to extinction. Most notable perhaps is its poisoning of the living system in which it was embedded, which was effective in rendering the population infertile (particularly during the dissemination of microplastics in the nutrients they consumed during the Plasticene, which spread all through their bodies and wrecked many vital functions, including their reproductive systems).

In addition to these various forms of self-enclosure, the other thing we're miserably failing to rhyme with the basic principles of life is what we've come to call the Two-Splitter Scheme: a peculiar mechanism—on which all of the above was dependent—by which everything was divided into and permanently bound to *just one of two parts within a binary pair*, e.g. male-female, human-nonhuman, reason-passion, mind-body, nature-culture, and so on. Baffling as its underlying logic may be, we are able to clearly distinguish a common effect: the subjection of one to the other (whereby which is which is determined by the former's [re]iteration of the latter as sub-ordinate or sub-human), accompanied by various forms of violence through repression, oppression, extorsion, and exploitation.

That's what set in motion the formation of "borderlands": vague and undetermined places created by the emotional residue of an unnatural boundary.

Commented PHV554 Report_Admin: Direct quote from p.35 of "Borderlands/La Frontera: The New Mestiza" (1987) by Gloria Evangelina Anzaldúa (American scholar of Chicana cultural theory, feminist theory, and queer theory).

It was in the final spasms of this paradigmatic complex that "the virus" (made synecdochal to "the pandemic") was *itself othered*: portrayed and treated as something that came from outside of the /sub/species-enclosed subjecthood's bounded selves; as an evil invader or bioterrorist that had to be combatted, controlled, and defeated. And not as something they'd themselves created through the machinery that was co-extensive to them. Not as something that *was them*, at least not in so far as its becoming-pandemic was literally and figuratively a mutation in the political ecology of global capitalism.>

Commented PHV554 Report_Admin: We've found a variation of this phrase in an article published in March 2020 by literary and political theorist Bram Ieven and philosopher and critical theorist Jan Overwijk, titled "We created this beast: The political ecology of COVID-19".

>Interestingly, the "facture/authorship" aspect that defined their notion of "culture" (as opposed to "nature") was overseen in this case.

So when the beast they had so laboriously created began releasing the stench of its decomposition and the curelessness of its unliveability became increasingly evident, their othering of "the virus" (sic. "pandemic")—i.e. their non-recognition that its existence was spun from their own toxic yarn and then back into the fabric of their ill-being—made their response capacity as poor as their incest-ridden genomes.

For example, they did things like pump huge amounts of money—which suddenly appeared could be coughed up in a heartbeat, for "an emergency", though it hadn't been for the environmental crisis (which is striking because we've found vast amounts of scientific data heavily supporting, by their own revered "confidence coefficient" standards, that they were *way beyond an emergency*)—into salvaging the broken parts of the broken machine, thereby destroying the best healers on the planet for the sake of saving an economy that was killing everyone.

And so the bucketloads of bullshit jobs- that were hollowing out peoples souls were preserved, while essential workers were buried deeper and deeper into the meritocratic, self-blame-inducing Debtocalypse of an economic system that was soaked in blood from the very beginning.

Commented PHV554 Report_Admin: This term was coined by American anthropologist, anarchist activist, and author David Graeber in his 2018 book "Bullshit Jobs: A Theory".

-While the term "bullshit job" appears only in reference to human persons, the data in the LL2 show that the largest part of these jobs were in fact reserved for plant- and animal-persons [see entry for "intensive farming/industrial agriculture").

So, of course, the existence of the interconnected being who remained bound to subsisting on this ailing machinery, and very large parts of whom provided the care- and reproductive work (sic. "free labour"; sic. "informal slavery") that was instrumental to that machinery's capacity to function, became *impossible*.

But there was no coalition because there was no recognition that it was fucked up for some of them in the same way that it was fucked up for the other some of them. And that it was killing them, too, however much more softly.

Commented PHV554 Report_Admin: This is a bastardization of a phrase on p.140 of a book titled "The Undercommons" (2013) by management and black studies scholars Stefano Harney and Fred Moten.

It got bad though.
People couldn't breathe in so many ways.

With the first outbreak of corona in 2020–21, it became clear that it wasn't possible to return to normality. Those who expressed a desire for such a thing were those who'd lost their heads upon sticking them in the sand of inborn privilege. No less for them though than for any other Gaian being, going back to normality was *the worst thing that could possibly happen*. Normality was deadly.

Among those who still wore their heads, it happened for the first time in living memory that people from across the planet common-*sensed* (rather than commonly knew) so many things that were crucial for the possibility of breathing life into an asphyxiating, asphyxiated world. And others could sense that others could sense it.

Interestingly, there is in the LL2 one piece of writing, which circulated widely, that describes the pandemic as a portal: a gateway between one world and the next, a rupture forcing humans to break with the past and imagine their world anew:

We can choose to walk through it, dragging the carcasses of our prejudice and hatred, our avarice, our data banks and dead ideas, our dead rivers and smoky skies behind us. Or we can walk through lightly, with little luggage, ready to imagine another world. And ready to fight for it.

Commented PHV554 Report_Admin: Direct quote from an article published in April 2020 by Indian novelist Arundhati Roy, titled "The pandemic is a portal".

We're quite sure the scenarios that were imagined in and through this text as constituting a "portal" were not quite of the same nature as those that actually happened.

In any case, the ones that were imaginable didn't happen.
The narrow timespace settings of the overbearing world were simply unfit for them.

After the first wave, the world was still burning, and that became the bellwether for what there was left to salvage in the second. We won't speak of the events that followed; detailed accounts can be found in the LL3.

Other events that were unimaginable, though, did happen—spurred on by a relay of breath that came from somewhere else; That seemed like it came out of nowhere.

Commented PHV554 Report_Admin: From Harney & Moten (see reference above).

Because heaving all the while under the thinning skin of the Earth's atmosphere had been a conspiracy of breathers, and an aspirational politics in which people had learned to *con-spire*, to breathe together. And to think conspiracy otherwise: not as a sinister association, but as the grounds for a livable politics.

Commented PHV554 Report_Admin: Anthropologist Natasha Myers writes this in reference to anthropologist Tim Choy, in an article from January 2020 titled "How to grow liveable worlds: Ten (not-so-easy) steps for life in the Planthroposcene".

Paraontological totality emerged from the mass of the constantly excluded pressing outside the gates of that in which they were forcibly included. It emerged for those who were not there, who danced passionately in the void of their subjectivity, who remained in the hold, in the break, as if entering again and again the broken world, in the incessant (dis)location of themselves, conspiring and becoming complicit in their own dispossession.

By no willful act or utterance—because one does not possess the kind of agency that can hold the regulatory forces of subjecthood—, but by allowing their subjectivity to be unlawfully overcome by others, they resounded within, through, and without their bodies, circulating in the synched-up borderlands of their impossibilitated existence —breathless but madly alive.

For ages they'd stayed illegible and unaffiliated, but now they learned—in being strung together—how to collude with each other's disaffiliation. A chain of recognition was created between the non-existing.

Commented PHV554 Report_Admin: This entire passage is a heavily bastardized version of several phrases from Harney & Moten's book (cited twice above).

As it locked, a surge of energy spun through the inter-yarn gaps of their fabric, gathering more and more energy in the null, in the vacuum within the void of the Anthropo-not-seen, of those who could not exist, who'd been held in limbo.

The hold didn't hold.

It never ceases to amaze how much dark energy can be held in a vacuum, and how explosive it can become when several points yoke to forge a loop. This one being among the more volatile, it took just a few revolutions for the whole thing to blow.

No words can describe the cataclysmic implosion that followed. That it caused the formation of a portal—The Cakewalk—should suffice to color the imagination.

Those who were in the loop were channeled through it, and they've been arriving ever since, finally finding refuge in a world that harbors many.

Given the ongoing arrivals, we know the portal is still active, and that it has since branched out to other Earth-based temporal zones. It's not visible of course, but it's certainly there and highly sensitized to the resonant frequencies that forged it, meaning it's receptive to any being who heaves in the void of her subjectivity—be it in extant, extinct, or yet unborn worlds.

Note to self:

We've noticed some of the pickneys around here wrapping our notes in chicle to shoot hoops at tunnel 4A, which stands close to the katabatic region around The Cakewalk openings, and which could on a windy day pull a ball in the reverse direction through the portal. Remember to give them proper balls and playfield.

While humans are still years away from viable life anywhere but Earth, the language and imagery used to describe this future venture mimics much of what is presented in science fiction, often maintaining existing Earthly political and social ideologies. Elon Musk, CEO of SpaceX—an aerospace company whose mission is to "make humanity multiplanetary"[1]—speaks of this project through the "colonization" of Mars. Aside from being dangerous and extremely costly, Musk's proposition to have a "life insurance for life as a whole" in the form of a backup planet in case "something terrible [...] happens on Earth, either made by humans or natural,"[2] presents a set of philosophical and ethical issues. Which Earthly lives are valued enough to be considered for outer space transport, and how would this travel operate? How would teleportation travel[3] and time travel affect bodily autonomy and corporeal subjectivity? While the popular imagination surrounding space travel, teleportation, and time travel has largely been shaped by mainstream works that feature white subjects and are often created by white male directors, thus mirroring Musk's own vision of Mars,[4] in 2019, three filmmakers, Meriem Bennani, Stan Douglas, and Jordan Peele, confronted the ways in which teleportation, quantum entanglement, and racial and national politics intersect. Through the films *Party on the CAPS*, *Doppelgänger*, and *Us*, Bennani, Douglas, and Peele, respectively, highlight that while science and technology purport a neutral approach that fosters a supposedly morally progressive path in technological advancement, they rather exacerbate existing inequalities, particularly as it affects those with racialized identities.

Working within the space of science fiction, Bennani, Douglas and Peele consider the dystopic potentials of technofetishism and the possibilities of continued scientific development in biotechnology in the "real world," researched and implemented to serve a white, Euro-American world order. While all three films are technically fictional, Douglas, and Peele envision alternative realities to our present moment, and Bennani speculates on a future in a manner that she refers to as "documentary," as the issues develop from current political policies and recent biological research.[5] In all three cases, humans continue on a technological trajectory that threatens the bounds of a discrete and unique corporeal self. Furthermore, the slippage, fractures, doubles, and mutations that arise out of teleportation and quantum entanglement challenge an understanding of biological citizenship, a marker of a national identity predicated on a wholly human, biological existence.

Peele's *Us* follows a home-invasion plot structure, in which the Black, upper-middle-class Wilson family is violently confronted with what appear to be their doppelgängers. As the narrative progresses, the origins and motives of their visual twins are revealed, and a larger political strategy based on biological manipulation is

1 "Mission," *SpaceX*, last modified June 10, 2020, https://www.spacex.com/mission.

2 Jackie Wattles, "Colonizing Mars Could Be Dangerous and Ridiculously Expensive. Elon Musk wants to do it anyway," *CNN*, September 8, 2020, https://edition.cnn.com/2020/09/08/tech/spacex-mars-profit-scn/index.html.

3 While physicists have been able to successfully achieve teleportation at the quantum scale of photons, transmitting the molecule's information rather than physical matter itself, teleportation at the scale of humans is not yet possible.

4 Some notable examples include *Star Trek* (1966), *The Fly* (1986), and the *X-Men* series.

5 Meriem Bennani, artist talk with Lisa Long, Julia Stoschek Collection, March 19, 2020, Berlin, YouTube, 51:18, https://www.youtube.com/watch?v=zZCzFLhvUuI.

6 "A Conversation between Meriem Bennani and Aria Dean," *Rhizome*, May 20, 2020, video, 45:01, https://vimeo.com/421611719.

7 "Mission," SpaceX. A prominent example of these stakes is the history and endurance of anti-Black medical racism. Throughout American history, Black and brown bodies have been the subject of unconsented or manipulated medical trials whose research would later serve a white public. The Tuskegee Syphilis Study is a prominent example. It was a clinical study conducted between 1932 and 1972 in which the United States Public Health Service observed the development of untreated syphilis in Black men who were told that they were receiving free health care. In another instance, the cancer cells of Henrietta Lacks, a Black woman, were kept and researched without her consent. These cells became the first immortalized human cell line. The HeLa cell line remains one of the most important cell lines in medical research.

8 Afrofuturism is a term coined by cultural critic Mark Dery in his 1993 essay

exposed as having instigated further inequality and instability. While the film is not directly about quantum mechanics, I will later argue that reading it through a quantum lens allows for a more nuanced understanding of the relationship between each Wilson family member and their respective doubles, as well as of the entanglement citizens experience on a collective and national level. While *Us* is a Hollywood blockbuster film, Bennani's *Party on the CAPS* and Douglas's *Doppelgänger* are video artworks. Bennani's eight-channel installation envisions a future in which teleportation has replaced airplanes as the dominant mode of long-distance travel. Conceptualized in 2017 while Bennani was researching subatomic teleportation and in response to Donald Trump's travel ban on citizens from the Muslim-majority countries of Iran, Iraq, Libya, Somalia, Sudan, Syria, and Yemen from entering the United States, *Party on the CAPS* speculates on the existence of an immigrant detention island in the Atlantic Ocean, the CAPS, run by Troopers (stand-ins for ICE) who intercept illegal African teleporters.

Bennani, a New York–based Moroccan artist, enlisted her family and friends in shooting the film, drawing from personal experiences to focus on the Moroccan community on the CAPS (short for half-way capsule, as residents are stuck between their place of origin and their destination). She envisions the project as a trilogy, in which the subsequent videos would focus on other North African countries.[6] *Doppelgänger* follows the journey of a lone Black female astronaut named Alice, who is teleported to the Hermes II spaceship on a distant planet as her double heads to another version of the Hermes II in a parallel reality. When Alice and her ship autopilot for unknown reasons and return home, she assumes her mission has failed. She actually has arrived, however, at a world in which everything from the direction of writing to the rotation of the sun is in reverse.

While the films employ varied strategies in visually and thematically approaching the issue of teleportation and quantum entanglement, all of them feature non-white and non-Euro-American characters. By creating speculative media landscapes in which parallel realities and futures are populated by subjects who often do not have access to the most advanced technologies, the filmmakers point to the true white-oriented stakes of sci-fi and real-world quantum projects that develop in the language of colonial visions for space.[7] For creative practitioners such as Camae Defstar, co-founder of Black Quantum Futurism (BQF), whose work operates within the space of a rich Afrofuturist[8] philosophical, historical, and methodological field that considers the teleportation of Black subjects, "the quantum physics side of Afrofuturism" becomes a space "to try and create a new practice, a new scientific theory."[9] Through this conception, the quantum world exists as a tool to depart from the structures of existing white-dominated politics,

"Black to the Future," following conversations with sociologist and technologist Alondra Nelson. However, scholars recognize Sun Ra's proposal of Afrofuturism to the music world in the 1950s and the coalescing of a cultural aesthetic. More broadly, Afrofuturism is a philosophy for art, history, and science that considers the experiences of members of the Africa Diaspora through science fiction and technoculture.

9 Laurent Fintoni, "Meet Camae Defstar, The Philly Activist Channelling Her City's Pain Through Music," *Fader*, December 18, 2015, https://www.thefader.com/2015/12/18/camae-defstar-mmgz-interview. Defstar and collaborator Rasheedah Phillips describe Black Quantum Futurism (BQF) as "a new approach to living and experiencing reality by way of the manipulation of space-time in order to see into possible futures, and/or collapse space-time into a desired future in order to bring about that future's reality. This vision and practice derives its facets, tenets, and qualities from quantum physics and Black/African cultural traditions of consciousness, time, and space. Under a BQF intersectional time orientation, the past and future are not cut off from the present—both dimensions have influence over the whole of our lives, who we are and who we become at any particular point in space-time. Through various writing, music, film, visual art, and creative research projects, BQF Collective also explores personal, cultural, familial, and communal cycles of experience, and solutions for transforming negative cycles into positive ones using artistic and wholistic methods of healing. Our work focuses on recovery, collection, and preservation of communal memories, histories, and stories." See "About," Black Quantum Futurism, https://www.blackquantumfuturism.com/about.

rather than as another strategy to uphold white-led and -incentivized systems of power.

In the essay "Black Timescapes, Time Travel + Temporal Displacement," Rasheedah Phillips, the other co-founder of BQF, complicates engagement with the quantum space and raises concerns with "traditional Eurocentric" depictions of "time travel/temporal displacement fiction."[10] In addition to issues of access to a "high tech Machine that only a privileged few can have access to," Phillips warns that, in traditional depictions, the "inventors in possession of the machine are somehow endowed with the wisdom and moral righteousness to know when, how, and what to use it for," and that the "time traveler is typically white male, often a scientist or astrophysicist," whose travel is "centered around his wants or needs and his traveling is allowed to disrupt reality."[11] Phillips also notes that women sci-fi "travelers are often traveling or are time-displaced involuntarily—they usually cannot control when and where it happens or where they go, or don't choose to have it happen. Usually men are in control of when and/or where they travel to. Black women time travelers depicted in film are exceedingly rare."[12]

Instances of Black characters, and particularly Black female characters, being subjected to unconsented or manipulated teleportation or quantum travel are also prevalent within a wider Afrofuturist literary tradition. Octavia E. Butler's novel *Kindred* (1979) is a notable example, in which Dana, a young Black woman, finds herself transported back and forth between her life in Los Angeles in the 1970s and a plantation in the Antebellum South, beyond her control and seemingly at random.[13] These issues of agency and consent that quantum teleportation brings up are explored more broadly in each film discussed here. In Peele's *Us*, it is only after Adelaide Wilson, the mother and protagonist, is locked inside the family's holiday home and shackled to her living room table that the true nature of the family's doubles are revealed. Red (Adelaide's double) explains that the doppelgängers are actually human clones called the "Tethered." While Red doesn't name who orchestrated the Tethered project, she explains that they were created as a means of controlling the human population. It is suggested that every person in the United States has a Tethered that they are seemingly unaware of and did not explicitly consent to having created.

Contained to live in an interconnected web of tunnels and former underground train stations beneath the surface of the United States, the Tethered mimic the movements of their corresponding humans in the world above. In the beginning of the film, this is expressed with a couple enjoying a snack on the Santa Cruz boardwalk as their respective Tethereds eat the bloody flesh of rabbits—the only food source in the dimly lit, dystopic world below. While it is not made clear whether this pantomime is a reflection of external conditioning or rather an internal drive akin to a

10 Rasheedah Phillips, "Black Timescapes, Time Travel + Temporal Displacement," *Black Quantum Futurism*, April 14, 2020, https://www.blackquantumfuturism.com/post/black-timescapes-time-travel-temporal-displacement.

11 Phillips, "Black Timescapes, Time Travel + Temporal Displacement."

12 Phillips, "Black Timescapes, Time Travel + Temporal Displacement."

13 Throughout the course of the novel, Dana realizes that she is travelling back in time on a seemingly fateful mission to protect one of her relatives. The violence of this quantum travel, taking place even without a machine, is made all the more dangerous as she, a Black woman, is brought back to her white ancestor's plantation in the South.

14 Natalie Wolchover, "The Universe Is as Spooky as Einstein Thought," *The Atlantic*, February 10, 2017, https://www.theatlantic.com/science/archive/2017/02/spooky-action-at-a-distance/516201.

15 Elena Howe, "Seven Directors 'Weaponize Storytelling' to Stand against Hatred and Violence," *Los Angeles Times*, November 30, 2017, https://www.latimes.com/entertainment/envelope/la-en-mn-directors-roundtable-20171129-story.html.

16 Alexander G. Weheliye, *Habeas Viscus: Racializing Assemblages, Biopolitics, and Black Feminist Theories of the Human* (Durham, NC: Duke University Press, 2014), 8.

certain personal entanglement, the Tethereds' actions always appear as insufficient. They lack tools as well as a physical capacity, and many speak in grunts and move in a haphazard manner. Red explains that the Tethered were abandoned when the experiment failed, that the creators "figured out how to make a copy of the body but not the soul. The soul remains one, shared by two."

While the Tethered are first presented as non-biological and potentially even nonhuman, I would argue that the relationship between the Tethered and the humans ontologically operates akin to quantum entanglement. Albert Einstein deemed entanglement "spooky action at a distance," referring to the phenomenon that occurs when two particles remain linked so that the observation and actions performed on one affects the other, no matter their distance.[14] The spookiness lies in physicists' uncertainty over how these particles operate and seemingly react to each other by, for example, spinning in opposite directions without any visible communication. Much like quantum particles, the Wilson family and their doppelgängers appear intrinsically linked yet inverted in abnormal action and, as suggested by Red, seem to share a soul, an internal anchor that alludes easy description from writers and physicists. As a child, Adelaide met Red, her Tethered, when Red escaped from the captivity of the underground tunnels and exited through a path in the Hall of Mirrors on the Santa Cruz boardwalk. The adult Adelaide seemingly blocked out any memory of this traumatic encounter; however, she expresses unease to her husband at the beginning of the film, recalling the feeling of being followed by the girl who she had seen so many years before. There is a flashback to a scene in which Adelaide, a professional dancer, reenacts a ballet routine as a duet between both of the women in the tunnels. Before performing together, they practice apart, though in both instances they move in perfect unison without spoken communication or planned choreography.

The architecture of the tunnels as underground holding vessels limits movement and opportunity and inhibits agency for the subjects held captive within them, thus operating as a physical manifestation of the hallucinatory "sunken place" in Peele's 2017 film *Get Out*. In *Get Out,* the "sunken place" is a metaphor for systems that oppress Black people, such as, as Peele has specifically identified, the prison-industrial complex and the lack of representation of Black people in film and in society more broadly.[15] When Gabe Wilson asks the Tethered, "What are you people?," he challenges their subjectivity and status, questioning whether they are fully human and enforcing the hierarchical ordering "of the Homo sapiens species into humans, not-quite-humans, and nonhumans" that Alexander G. Weheliye, professor of African American Studies at Northwestern University, has argued is dictated by the fact that all of the "modern human[s]" are subject to racialization.[16] In *Us*'s parallel reality, while biologically existing as humans, the Tethered are denigrated to a "nonhuman" class. Red's reply, "We're Americans," implicates a distinct national identity as a precursor toward being recognized as fully "human."

The "not-quite-humans" and "nonhumans" framework for conceptualizing racialized subjects is directly addressed in Douglas's *Doppelgänger.* When the astronaut Alice unexpectedly returns to Mission Control in one universe, she is treated with care and concern, offered vitamins and rest, while in the other she is viewed as a suspect and potential alien and is subsequently quarantined. As the only Black character in the film, the comment by a scientist that "whites like to think they belong in the US, but every time they see an Indian they are reminded they descended from settlers, so they fantasize an alien more alien than themselves" in response to the incredulity of Alice being an alien implicates race within the discussion of a foreign or immigrant other. To construct an alien is to deflect historical guilt and maintain whiteness as the status quo. The irony of creating an othered subject is that this being exists as a projection—"we" are the alien. This metaphor is most directly expressed by the title of Peele's film, *Us*, or rather the U.S., in which the Tethered present as dark shadows of the mirror image of each American, carrying the burden of past violence and historical memory.

Weheliye's argument that race operates "not as a biological or cultural classification but as a set of sociopolitical processes of differentiation and hierarchization, which are projected onto the putatively biological human body," is useful in considering the effects of teleportation on the bodies in Bennani's *Party on the CAPS*.[17] For Weheliye, "racializing assemblages" represent "the visual modalities in which dehumanization is practiced and lived," and the racialized biological component of "human flesh" becomes an anchoring mechanism by which to maintain biopolitical control.[18] A body is intrinsic to forming an identity and also crucial for governments in dividing and manipulating individuals into demarcated groups; however, in Bennani's speculation of the future, a sealed body is not something guaranteed. For the North Africans who illegally teleport to America and are intercepted by Troopers en route, some bodies are not able to be wholly reassembled. In extreme instances, refugees are reduced down to ethereal matter, darkly pictured by a pair of shoes walking by themselves. For many, the violation at the quantum level results in "mega ear" and "plastic face syndrome," side effects of the violence of having one's body intercepted by the state mid-travel. Even more problematic explains ZIP, a quantum smuggler, these conditions are often passed down to future generations. To rid themselves of these corrupted bodies, some residents enlist ZIP to teleport them into America to take over an existing body. To have one's own body in reassembled condition on the CAPS is a privilege itself, but to have the funds to escape the island with

17 Weheliye, *Habeas Viscus*, 5,6.

18 Weheliye, *Habeas Viscus*.

19 Stefano Harney and Fred Moten, *The Undercommons: Fugitive Planning & Black Study* (Brooklyn, NY: Autonomedia, 2013), 96.

20 Ruha Benjamin, "Informed Refusal: Toward a Justice-based Bioethics," in "Resisting Power, Retooling Justice: Promises of Feminist Postcolonial Technosciences," ed. Anne Pollock and Banu Subramaniam, special issue, *Science, Technology, & Human Values* 41, no. 6 (November 2016): 968.

21 Benjamin, "Informed Refusal," 968.

one's own body is a luxury for the few with money. Due to the risk of corporeal incohesion and maintenance, Fiona, the animated crocodile narrator, explains that the CAPS "is a place where the idea of having a body itself is never taken for granted."

On the diasporic site of the CAPS, the quantum intervention of teleportation has highlighted an inherent tension between biological and political framings of a self, in which an individual's corporeal condition is at risk. What does an attack against a body at the quantum level mean for personal subjectivity and identity, particularly when an individual is targeted for their racial or national position? If having a recognizable human body is necessary for political participation, to what extent can a body which is fractured exist within the normative communal white-space, and to what end do individuals have a right or ability to object to engagement with structures of oppression?

Those living on the CAPS practice a strategy of refusal, or what Fred Moten and Stefano Harney define as "the refusal to be refused."[19] In response to a techno-powered global police state that surveils and criminalizes refugees, the residents of the CAPS employ music and celebration as tools with which to challenge the conditions as they have been presented and to imagine a new life within an impossible environment—pleasure as biopolitical resistance. In addition to unplanned body morphs as part of interrupted travel, some residents intentionally undergo "rejuvenation" surgeries to reverse ageing by decades and take control of their appearances. In this sense, the process of being reconstituted poorly and arguably as defective presents as an emancipatory proposition. If citizenship is predicated on biological appearance, then the decision to metamorphosize the self, even to the point of unrecognition, becomes an act of "bio-defection." This is best evidenced by a woman celebrating her eightieth birthday who appears to be in her mid-thirties. As presented by the sociologist Ruha Benjamin, biodefectors are "those who attempt to resist technoscientific conscription" and "resist biologically based citizenship."[20] While Benjamin cites genetic ancestry testing and screenings for breast cancer genetic mutations among procedures that enact biocitizenship, biocitizenship more broadly is a "family of concepts that sets out to understand the rights and responsibilities that accompany enhanced techniques for intervening in life itself."[21]

Considering the biological and technoscientific processes involved in the theoretical acts of teleportation and quantum entanglement, the concept of biocitizenship is useful for identifying strategies these filmmakers take in the fictive sense and for imagining new possibilities as viewers in identifying, challenging, and even refusing technologies that instigate racial and nationalist difference while claiming neutrality. At a time of revived discussions around identity politics in creative practices, Bennani, Douglas, and Peele's works fulfill a need for expanded representation of both creators and subjects without self-tokenizing or pigeonholing the narrative. The presence of non-white characters within the quantum space works to implicate American society as an enforcer of white supremacist structures that enact violence on racialized bodies. These works focus on non-white subjects, but, at the same time, they point toward the actions of the oppressor. As viewers, we are invited to consider and observe who has and continues to commit violence through strategies of segregation, containment, and even physical assault to the point of death.

In response to the shooting of Jacob Blake at the hands of the Kenosha Police Department in August 2020 and a continued resurgence in national and international protests in support of the Black Lives Matter movement following the murder of George Floyd, the artist Ajay Kurian posted on Instagram, "if the result of a quantum experiment is observer-bound, then perhaps because the quantum project of America is being newly observed, the experiment this time may turn out different."[22] On a national level, the actions of police officers, and of white Americans more broadly, are being scrutinized with urgent criticality. Within the cinematic space of these three films, the considerations of the Tethered's creators, Alice's colleagues, and the American Troopers offer a strategy for imagining new ways of understanding the violence of the American quantum project. These films are commentaries on Euro-American immigration policies, surveillance, state violence, and the consequences of a complicit and often ignorant white majority in the context of the observer effect, but perhaps they also suggest the potential for culture to change the grounding of reality.

For the racialized bodies in these films, the state and the American majority enact a biological violence that is felt ontologically as well as corporeally. As the host of the birthday party in *Party on the CAPS*, played by Bennani's mother, explains, "They don't understand our bodies. Because when you have teleported and been intercepted, your body is affected, it never feels the same again." For all bodies that are racialized and othered through American political mechanisms, there is a willful lack of awareness from those in power of the emotional and physical repercussions of these interventions. There is also a limited understanding of what constitutes a body and what bodies are recognized as politically viable. Through these films, the artists highlight the violence of doubling, splitting, manipulating, quarantining, and surveilling a body. To the extent that they use human cloning and teleportation, the theoretical technologies of quantum modulation on the human body, as foils to interrogate value-systems, the artists discussed present a vision of the future that cannot be divorced from the conditions of reality. If we want to avoid the "terrible" Earthly event that Musk fears and that has inspired him to plan a terraforming of Mars, we need to recognize that Earth in its current form can

22 In physics, the observer effect is the disturbance of an observed system by the act of observation—to disturb something provokes the system to change. More recently, scientists are considering the possibility that space-time might not be the most fundamental part of existence and that something like consciousness might be.

be a terrible place for those who are not considered valuable by the state and others in positions of power. If alternative forms of space travel become possible, then in order for Mars to not fail in ways that the Earth has, society will need to realign racist value systems and break the current mentality of colonial and imperial rule. As the artists here visualize, an alternative present or future space will look and operate akin to our own if we don’t change anything now.

COLORS
Sexy British woman's voice

electronic music (JOAKIM'S BEST OF MUSIC YOU CAN DANCE TO IN 2014?) played throughout, with hotline dialogue interspersed

pressing buttons makes "hero" from Mac sounds, a short cry, a kid's toy, etc.

SplendidIsolationMenu: Hello, and thank you for calling Splendid Isolation.

pause

Tell me, given the choice of anyone in the world, whom would you want as a dinner guest?

pause for several seconds

Well?

pause for several seconds

Would you like to be famous? In what way?

pause for several seconds

Before making a telephone call, do you ever rehearse what you are going to say?

pause for several seconds

Why?

pause for several seconds

What would constitute a "perfect" day for you?

pause for 7 seconds

Go on.

pause for 14 seconds

When did you last sing to yourself? Or to someone else?

pause for several seconds

If you were able to live to the age of 90 and retain either the mind or body of a 30-year-old for the last 60 years of your life, which would you want?

pause for several seconds

Do you have a secret hunch about how you will die?

pause for several seconds

Yeah?

pause for several seconds

Name three things you and I seem to have in common.

pause for 15 seconds

For what in your life do you feel most grateful?

pause for 7 seconds

If you could change anything about the way you were raised, what would it be?

pause for 7 seconds

Take four minutes and tell me your life story in as much detail as possible.

pause for four minutes

If you could wake up tomorrow having gained any one quality or ability, what would it be?

pause for several seconds

If a crystal ball could tell you the truth about yourself, your life, the future or anything else, what would you want to know?

pause for 5 seconds

Yeah?

pause for several seconds

Is there something that you've dreamed of doing for a long time?

pause for several seconds

Why haven't you done it?

pause for several seconds

What's the greatest accomplishment of your life?

pause for several seconds

What do you value most in a friendship?

pause for several seconds

What is your most treasured memory?

pause for a minute

What is your most terrible memory?

pause for a minute

If you knew that in one year you would die suddenly, would you change anything about the way you are now living?

pause for several seconds

Why?

pause for several seconds

What does friendship mean to you?

pause for 10 seconds

What roles do love and affection play in your life?

pause for 10 seconds

Tell me something you consider to be one of my positive characteristics.

pause for several seconds

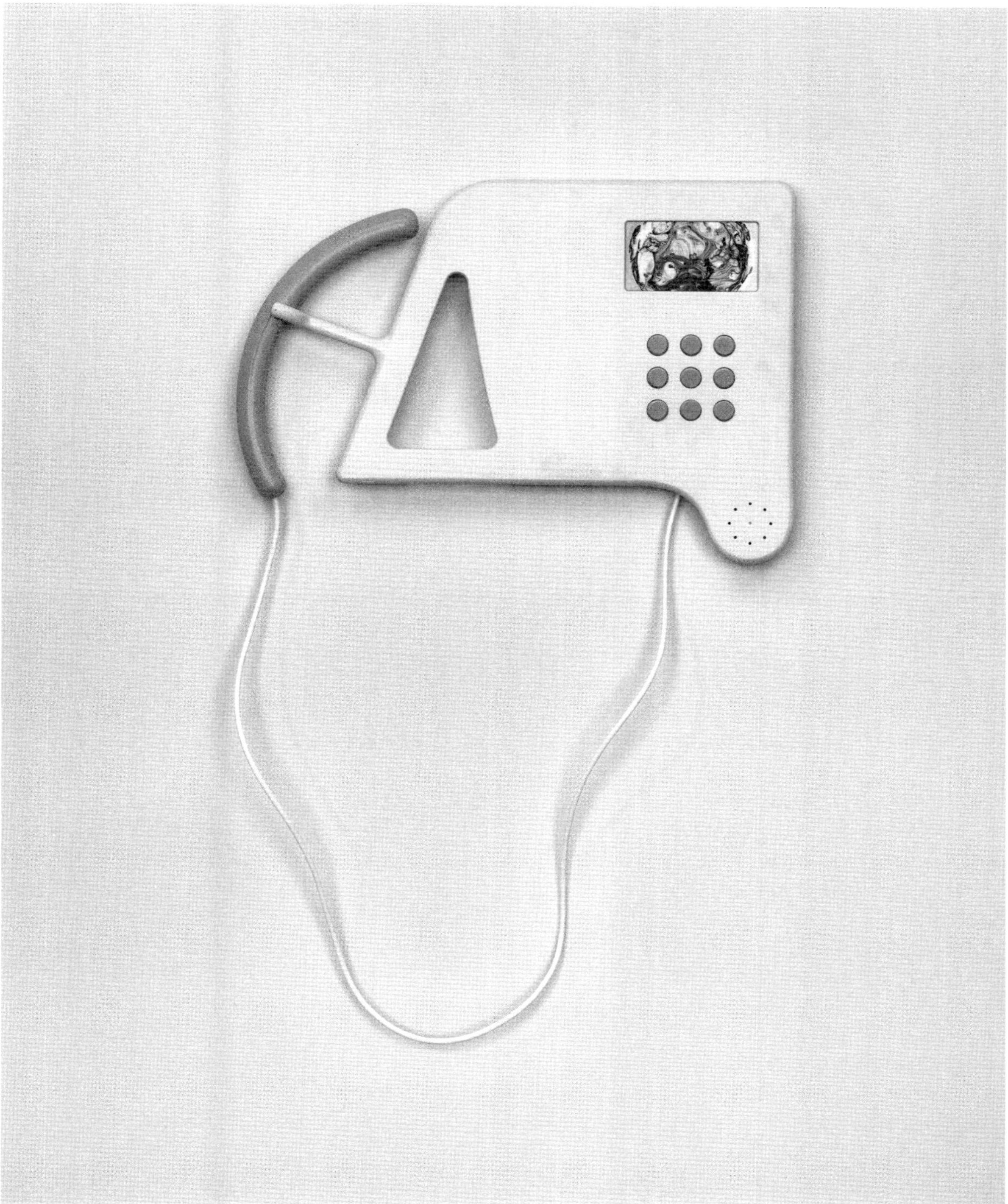

Camille Henrot, *Bad Dad & Beyond*, 2015, from the series *Interphones*. Interactive sculpture, mixed media

giggle Well, I think you're cute...You go again.

pause for several seconds

You seem very bright...Your turn.

pause for several seconds

You've got a weird sense of humor. But I like it. Go again.

pause for several seconds

Well, you're adventurous. I mean, you're here, right? Last turn.

pause for several seconds

I like your capacity to feel.

pause for a couple seconds

How close and warm is your family? Do you feel your childhood was happier than most other people's?

pause for 10 seconds

How do you feel about your relationship with your mother?

pause for 10 seconds

Make three true "we" statements about the two of us. For instance, "We are both in this room feeling silly, but intrigued." You go ahead now.

pause for several seconds

Okay, I'll go. We might want to continue spending time together…

pause for 10 seconds

Hmmm...We...haven't had a connection like this in a while...

pause for several seconds

Complete this sentence: "I wish I had someone with whom I could share..."

pause for several seconds

If you were going to become close friends with me, please share what would be important for me to know.

pause for several seconds

Uh-huh

pause for several seconds

Tell me what you like about me; be very honest this time, saying things that you might not say to someone you've just met.

pause for several seconds

Okay...

pause for several seconds

Share an embarrassing moment in your life.

pause for 10 seconds

When did you last cry in front of another person?

pause for several seconds

And by yourself?

pause for several seconds

Tell me something that you like about me already.

pause for several seconds

What, if anything, is too serious to be joked about?

pause for 7 seconds

If you were to die this evening with no opportunity to communicate with anyone, what would you most regret not having told someone?

pause for 10 seconds

Why haven't you told them yet?

pause for 7 seconds

Your house, containing everything you own, catches fire. After saving your loved ones and pets, you have time to safely make a final dash to save any one item. What would it be and why?

pause for 10 seconds

Of all the people in your family, whose death would you find most disturbing? Why?

pause for 10 seconds

Share a personal problem and ask for my advice on how I might handle it. Also, ask your me to reflect back to you how you seem to be feeling about the problem you have chosen.

pause for several seconds

Uh-huh...

pause for several seconds

Uh-huh

pause for 10 seconds

Hmm. You know, I don't want to come off too strong, but I've gone through that stuff myself. I think Ultima Thule Tele-therapy is really great.

pause for several seconds

Maybe we can talk again soon?

pause for several seconds

End

In the face of Trumpism, the usual modes of criticism often seem inadequate, almost voided by the sheer scale of the calamity as well as its protean mix of the buffoonish and the lethal. If farce comes after tragedy, what comes after farce, and how do we respond to whatever that is?

One way to approach the predicament is to push the old critiques to a new level. "Never before has an age been so *informed* about itself," Siegfried Kracauer wrote in 1927 about the new visual culture of Weimar Germany, and "never before has a period *known* so little about itself."[1] This paradoxical condition—whereby superabundant information undermines actual knowledge—is heightened in our media environment, which overwhelms us with data even as it deskills us in interpretation, which connects us even as it untethers us. Another example, which also concerns the Weimar period, can be extracted from *Critique of Cynical Reason* (1983) by Peter Sloterdijk (written before he became a darling of the German Right).[2] According to Sloterdijk, cynical reason has the structure of fetishistic disavowal ("I know X is not true, but I believe it nonetheless"), and he offers the notion as a way to account for the blind eye turned to Nazism in its emergence in the 1920s. If we are to comprehend the Trumpist mentality, we might also need to raise this concept to a higher power, for cynical reason today does not care to know, or if it does know (and most Trump supporters do), it does not care. As the saying goes, critics take Trump literally but not seriously, while supporters take him seriously but not literally. Noncynical unreason, anyone?[3]

Certainly, a posttruth politics is a massive problem, but so too is a postshame society. How to belittle a leader who cannot be embarrassed? Or to mock one who thrives on the absurd? How to out-Dada President Ubu? And should we aim to add outrage to a media economy that feeds on the same?

Thoughts about a postshame condition lead one to stories about a preshame time, the most outlandish of which features "the primal father." In *Totem and Taboo* (1913), Freud derives this figure from the idea of "the primal horde" in Darwin, a great band of brothers ruled by an all-powerful patriarch. This awful father enjoys all the women in the horde (that is the only role women have in this far-fetched tale) and leaves the sons out sexually, to the point where they finally rise up, kill, and devour the tyrant. However, this act plunges them into guilt, and so they elevate the dead father again, now as a god or at least a totem, around which taboos are established (the taboos against murder and incest above all). Thus, for Freud, does society begin.

There is a way to read this fable of prehistory historically, as a gloss on the bourgeois revolutions that overturned the kings, that is, as an allegory of democracy, "the transformation of the paternal horde," as Freud puts it, "into a community of brothers."[4] Freud brings back the primal father several years later in *Group Psychology and the Analysis of the Ego* (1921), and, if *Totem and Taboo* reflects

1 Siegfried Kracauer, *Mass Ornament*, ed. and trans. Thomas Y. Levin (Cambridge, MA: Harvard University Press, 1995), 58.

2 See Peter Sloterdijk, *Critique of Cynical Reason*, trans. Michael Eldred (Minneapolis: University of Minnesota Press, 1987).

3 This saying is credited to different people, such as the Trumpists Peter Thiel and Corey Lewandowski, as well as the journalist Salena Zito.

4 Sigmund Freud, *Group Psychology and the Analysis of the Ego*, trans. James Strachey (New York: W.W. Norton, 1959), 54. Rightist personalities such as Milo Yiannapoulos like to call Trump "Daddy."

5 Freud, *Group Psychology*, 59. In a short text titled "Toward a Real Revolution" (1936), Georges Bataille writes: "Under autocracy, it is authority which grows intolerable. In democracy, it is the absence of authority." *October* 36 (Spring 1986): 35.

6 My brief analysis leaves out a huge piece of the electoral puzzle—why a majority of white women voted for Trump. But then they may also be subject to "the erotic tie," or, rather, subjected to it.

7 Robert Smithson, "Response to a Questionnaire from Irving Sandler" (1966), in *The Collected Writings*, ed. Jack Flam (Berkeley: University of California Press, 1996), 329.

indirectly on democracy, *Group Psychology* does the same with fascism—that is, with the return, long after the democratic decapitation of the king, of the figure of the dictator. Indeed, for Freud, the mass politics of the time induce a regression to "the group psychology of the horde": "What is thus awakened," he writes, "is the idea of a paramount and dangerous personality toward whom only a passive-masochistic attitude is possible." "The leader of the group," Freud concludes, "is still the dreaded primal father; the group still wishes to be governed by unrestricted force; it has an extreme passion for authority."[5]

Why recall the primal father in relation to Trump? It is risky to psychologize any group of people, let alone millions of voters, to totalize them in this way, but there is a psychic dimension to his support that we must come to terms with. No doubt many of his voters—and remember that he received 63 percent of the white male vote in 2016—are sexist and racist, whether secretly or not; certainly, most are angry at elites, too. But they are also—they are primarily—*excited* by Trump, excited to support him: there is positive passion here, not just negative resentment. It may be difficult for progressives to see why, but one way is to suggest that he tapped into "the erotic tie" that binds the horde to the primal father. For this figure both embodies the law (he lords it over the brothers) and performs its transgression (he can grope any woman).[6] A potent double identification opens up: the brothers submit to the father as authority and envy him as outlaw. And so, we have a celebrity president ("when you're a star you can do anything") as throwback primal father (the bully-in-chief), and there are legions of white guys who want to be his apprentices.

What is the Left to do when the Right appropriates its cultural strategies? The defanging of transgression is a familiar story: on one side, it is recouped as spectacle (think of the performance artist Marina Abramović); on another, it is incorporated in the art institution (like the leopards in Kafka that break into the temple only to become part of the ritual); on still another, it is dulled by repetition. (Already in 1966 Robert Smithson could speak of "the avant-garde of dissipated scandals.")[7] Yet the coup de grâce has come only recently, and it occurred in the political realm, where transgression was usurped by the Right. When Steve Bannon, a self-proclaimed student of Lenin, called for "the deconstruction" of the government, Trump responded with executive orders, appointments, and tweets that, with extraordinary ease, trashed all kinds of laws, customs, and conventions, some of which actually protected people without power. Fifty years after the Leninist cry "Smash the State!" was revived, the White House has done this work effectively on its own. That many Americans look to the FBI and the CIA for salvation is another sign of how upside down the situation has become.

Where does this predicament leave artists and critics on the Left? One measure is offered in the 2017 film *The Square* by the Swedish director Ruben Östlund. *The Square* hits the mark in two scenes in particular, which together reveal an art world torn between transgressive routine on the one side and ethical vigilance on the other. The first scene is an elegant dinner after a museum opening where a performance is given by a shirtless man pumped up like the Hulk in mid-metamorphosis. He cavorts around the banquet hall like an angry ape, and the culturati are amused, titillated, even provoked, but then they know the rules of the game—this is only a performance after all. Soon, however, the situation gets out of hand. The apish man mocks the artist, challenges him, then forces him to flee the room. The curator declares the performance over, but the grunting man carries on, molesting other attendees, then assaulting a woman. This display of aggression run amok upends more than cultural decorum. For a moment, the beast has become sovereign, and, though the film was shot before Trump was elected, the scene reads as an allegory of his transgressive rule. How to respond? No one in the high caste at the dinner seems to know, until, at last, the tuxedoed men, now crazed, rise up against this primal figure.[8]

The second scene is a press conference where the artist is interviewed by a female curator. An unseen man in the audience interrupts the conversation with verbal abuse that is directed mostly at the woman. As the disturbance continues, other audience members shout at him to shut up. Although his interruptions are obscene, they are also compulsive, and eventually it is explained that the man has Tourette's. Suddenly, the situation is reversed: now he is the victim, calls for tolerance ring out, and the ones who challenged him a minute ago are disparaged in turn. Another tense situation, a different cultural-political allegory, but with the same question: How to respond?

The brutish performance is transgressive. Sometimes the creaturely signals a crisis in the social order, and often violence erupts when law is suspended, as it does here.[9] The task then is not simply to restore the law, as transgression usually does with taboo, but to reinscribe it differently, not simply to clamp down on disorder, but to bend its disruption to a new configuration of community. That is the opportunity in the present period of political chaos: to turn disruptive emergency into structural change.

For its part, the tolerant press conference is a learning-play in what we might call "ethical aesthetics." Not long ago "relational aesthetics" appeared on the scene to suggest, in effect, that if sociality was largely wrecked by neoliberalism (recall the Thatcherite motto "there is no such thing as society"), then a remedial preserve might be carved out in cultural institutions: one could still interact with others there. The current case of ethical aesthetics is similar. As codes of conduct are

8 What happens when the sovereign *is* the beast, or the king becomes the fool? (The opposite case is a moment of carnival within the court that upsets its order, only then to reestablish it.) Might Trump inadvertently de-sovereignize the presidency in the process? And as the giant balloon figure of Trump suggests, might Père Trump double as Bébé Donald? (Recall that Senator Bob Corker described the Trump White House as an "infant daycare center.")

9 See Hal Foster, "Wild Things," in *What Comes after Farce?* (London: Verso, 2020), 23–31.

10 See MTL Collective, "From Institutional Critique to Institutional Liberation? A Decolonial Perspective on the Crises of Contemporary Art," *October* 165 (Summer 2018): 192–227. This vigilance can take on a minatory cast, as it does in the brilliant story "Now More than Ever" by Zadie Smith, *New Yorker*, July 23, 2018.

11 Here is an extract from the "J20 Art Strike" proclamation concerning the art strike on Inauguration Day 2017 that still holds today: "Despite its contradictions, the art world has significant amounts of capital—material, social, and cultural—at its disposal. The time has come to imagine and to implement ways of redirecting these resources in solidarity with broader social movements leading the way in the fight against Trumpism." The full text is reprinted in *October* 159 (Winter 2017): 143–47.

trashed in profession after profession, it becomes incumbent on cultural institutions to insist on them all the more, to be exemplary in this respect. Yet there is a problem here too: just as the social was in part outsourced to culture in relational aesthetics, so the ethical is now in part relocated there in ethical aesthetics.[10] These institutions cannot be treated as merely compensatory in this way. As we know well, the culture industry has its own share of apish Trumpets, and many of its boards are deeply implicated in the political economy of the Trump brand. This complicity has provided important leverage for renewed feminist and institutional critiques—from broad movements like #metoo to focused groups such as Occupy Museums and Decolonize This Place.[11] At the same time, we should not let bad actors get a free pass where they should be most held to account—in the political realm. The task is to demand it there as well, there above all.

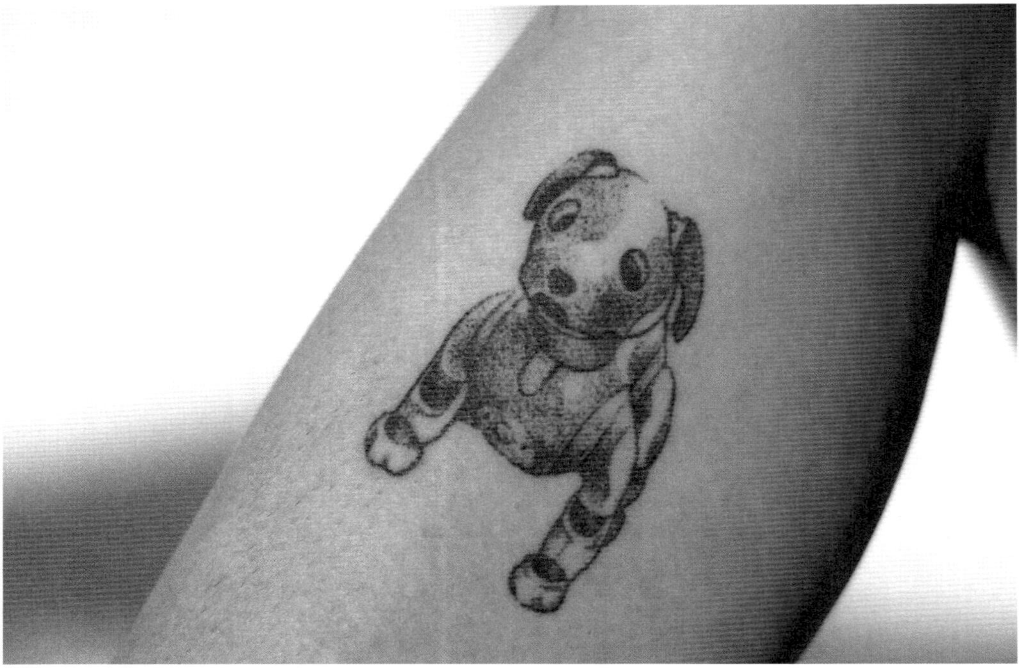

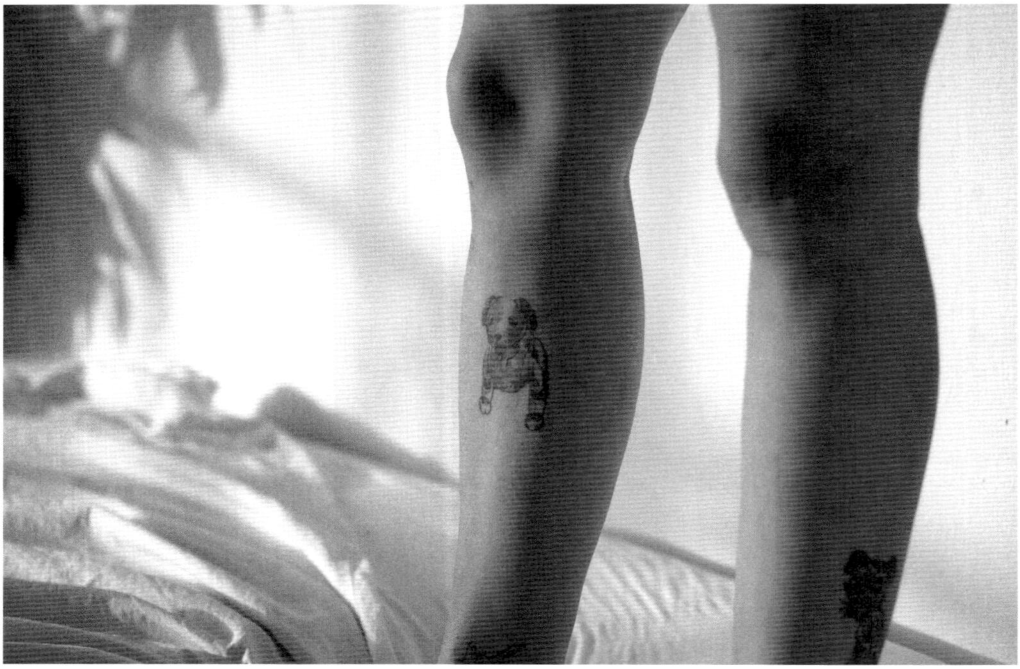

Tobias Kaspar, *Aibo*. Tattooed by Ivan Mitrović on Cassidy Toner's leg, photograph by Marie Matusz

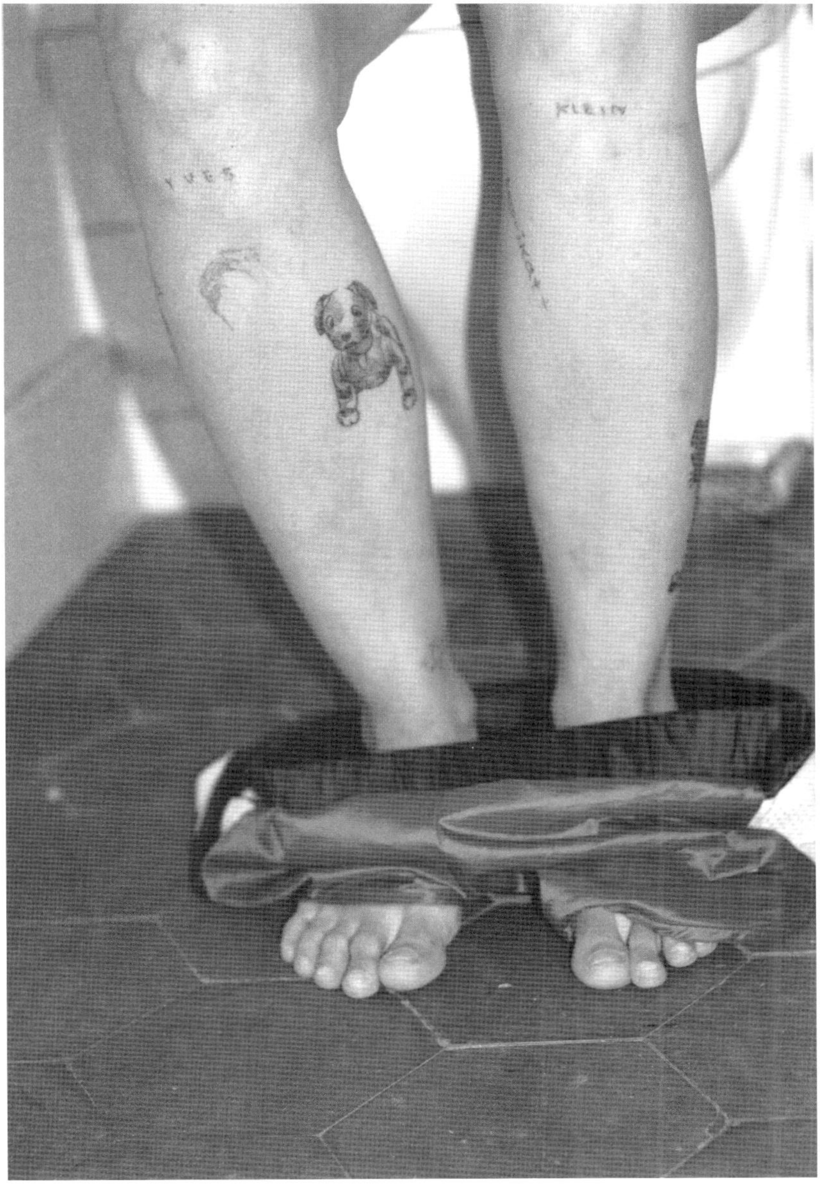
YVES
KLEIN

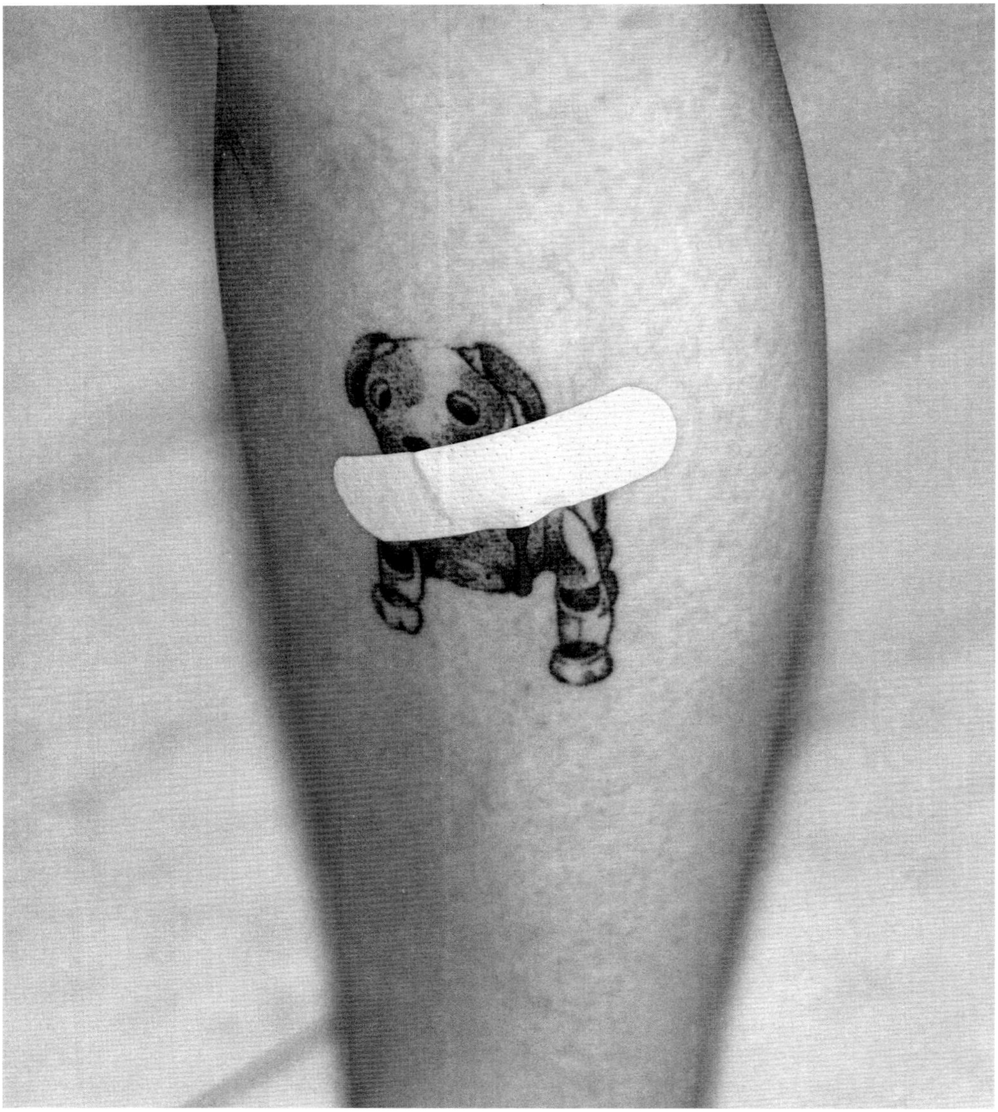

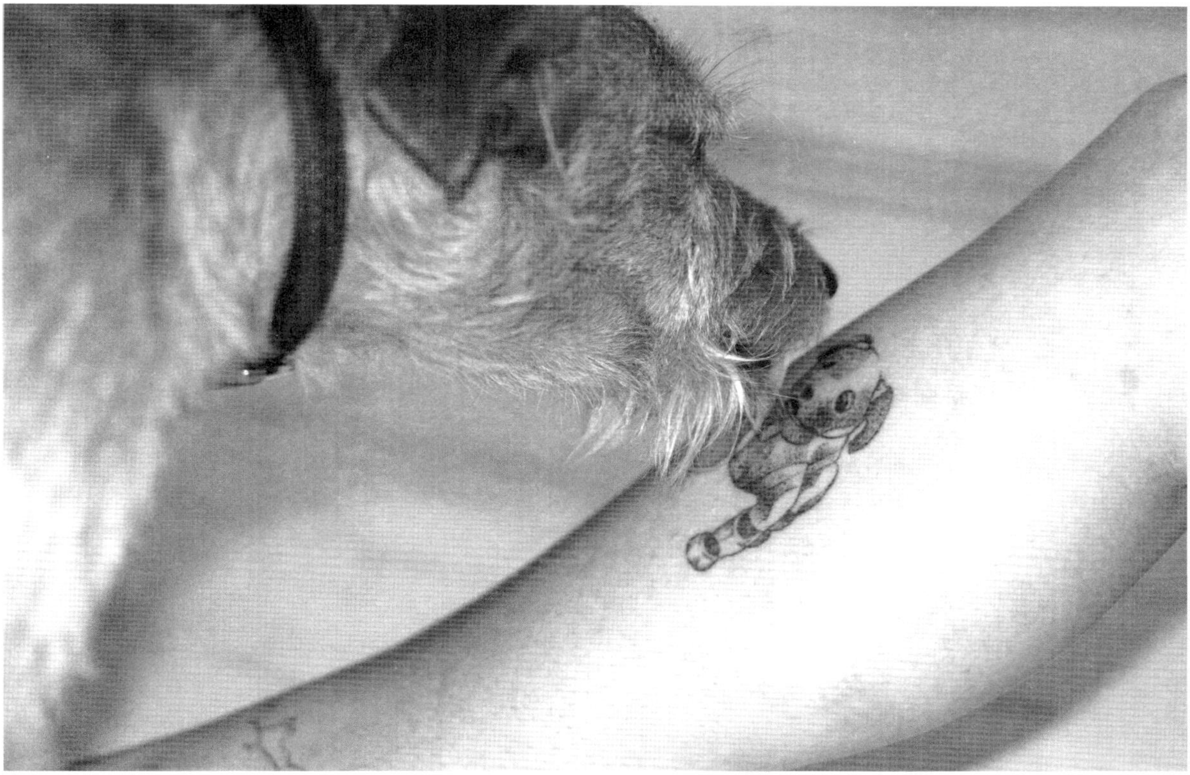

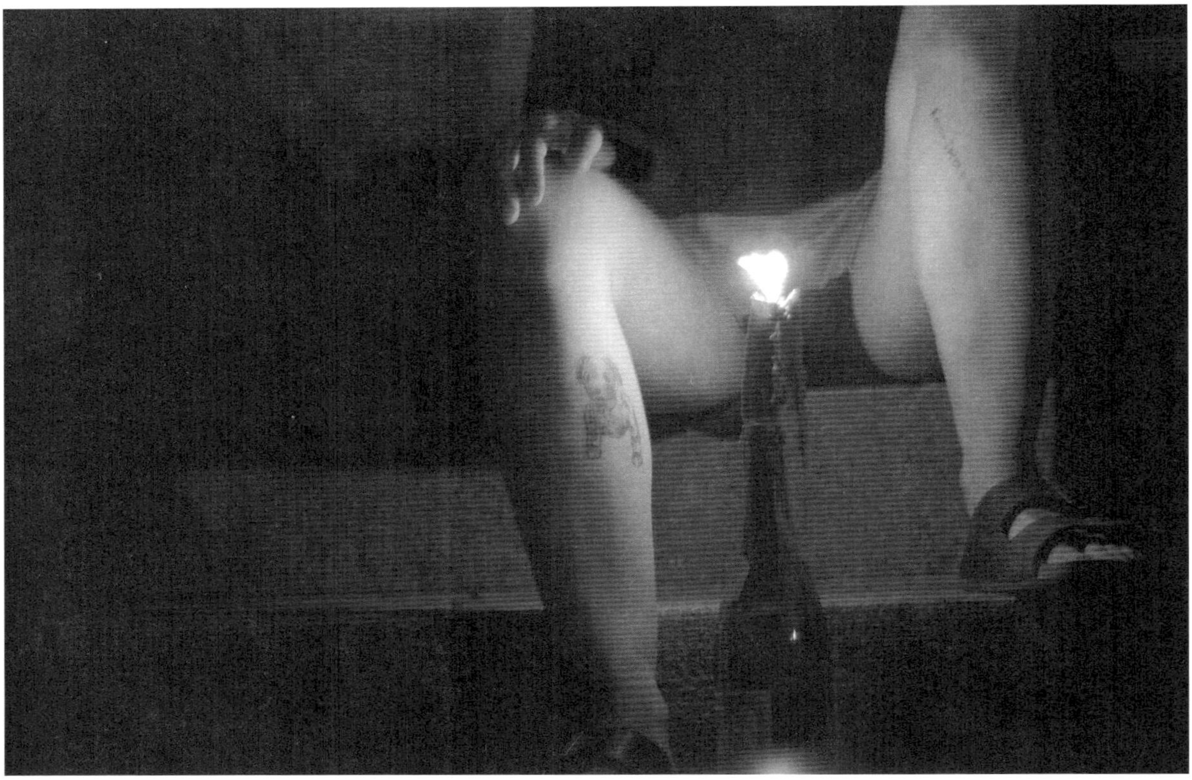

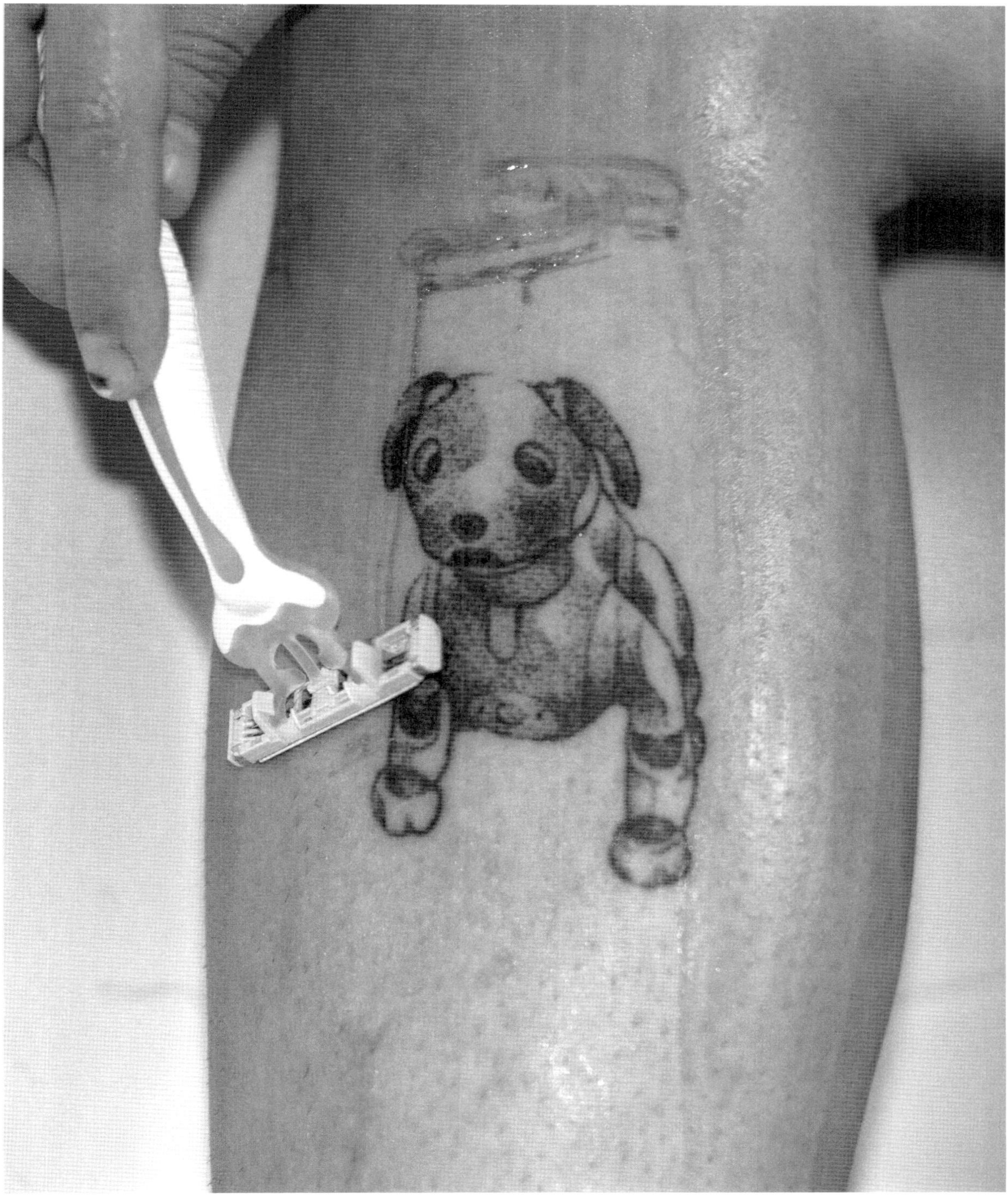

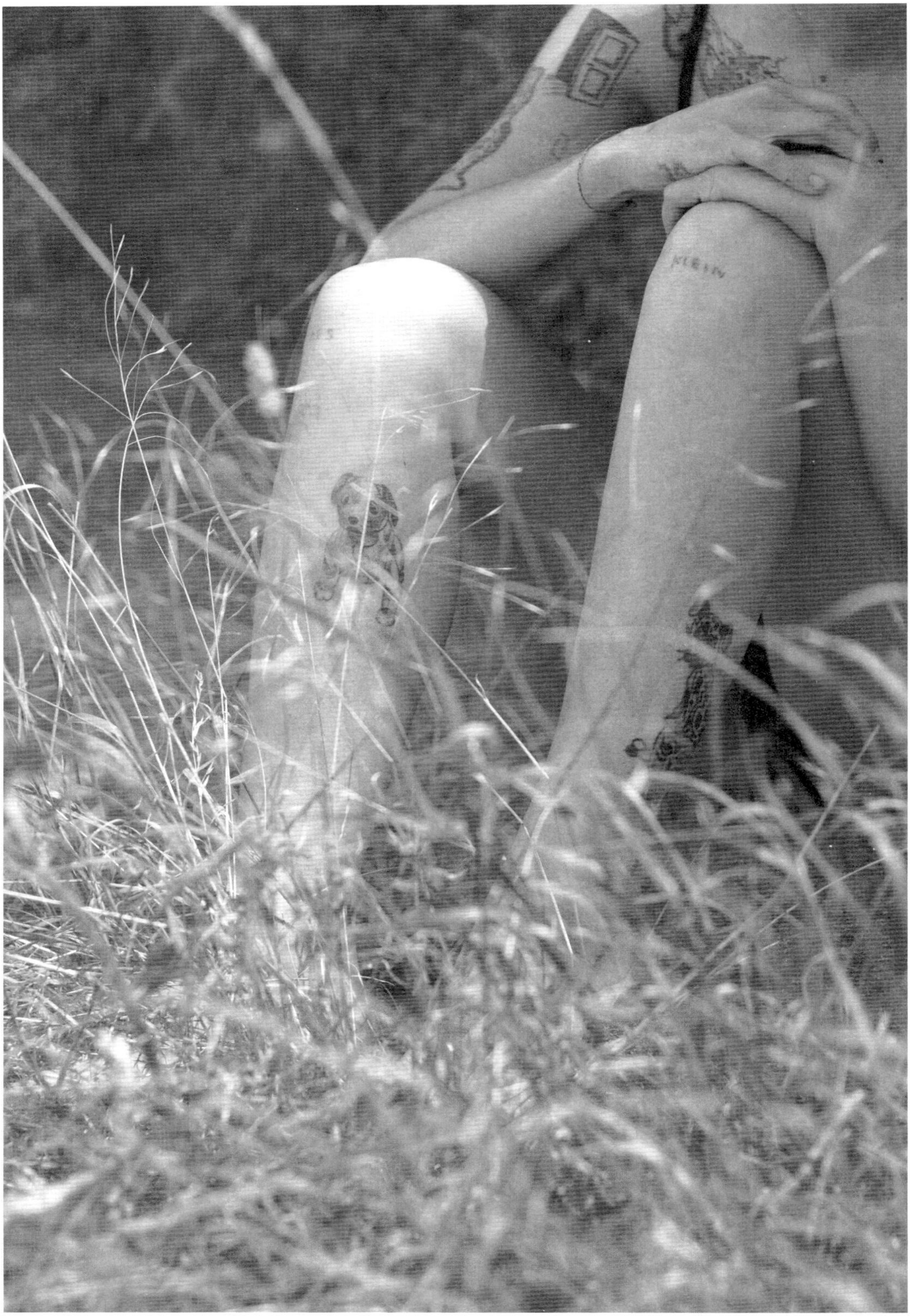

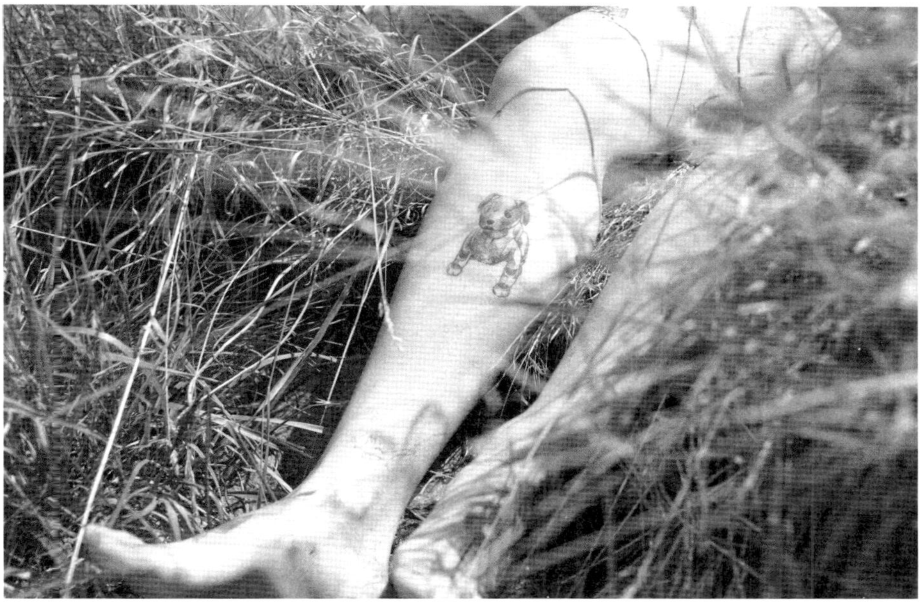

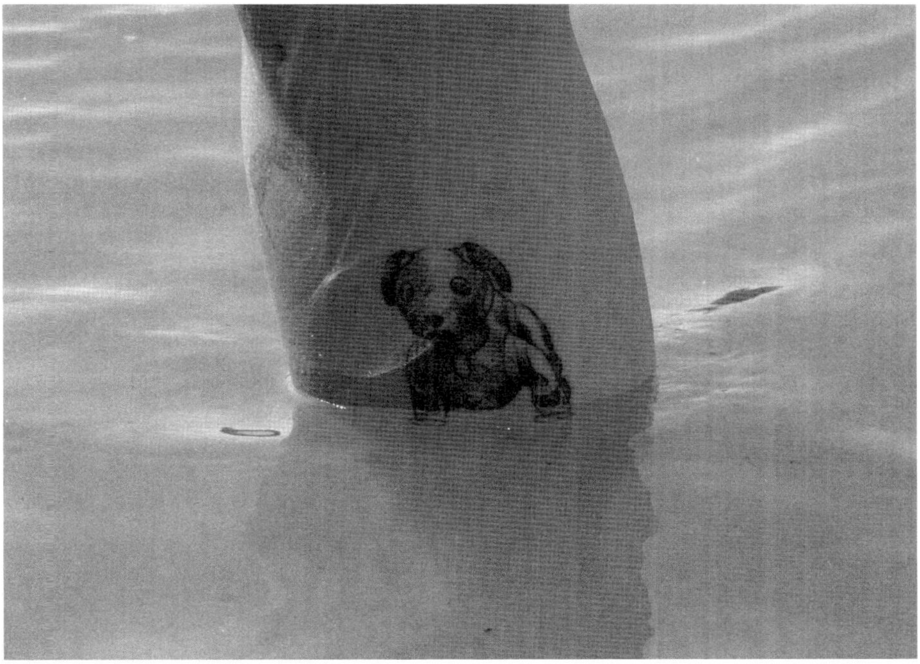

In this conversation, Frederick Cruz Nowell (musicologist) and X Zhu-Nowell (curator) talk with artists Devin Kenny, Dora Budor, and Cecilia Bengolea about recent projects that embrace variable subjectivities, listening relationships, and the fluidity of environments as constitutive elements of their art-making practices. Touching upon a range of topics—including instrumentality, musical "scratching," organs, and bodies floating in bodies of water—this conversation seeks to highlight the different ways these artists place the fluid and social experience of knowledge at the center of artistic contemplation.

Devin Kenney

FREDERICK CRUZ NOWELL: I was perusing through your music recently and came across a provocative remix that you made quite early in your career, about a decade ago. The track is a remix of the band Soundgarden's grunge rock earworm "Black Hole Sun" from the mid-nineties, which you "scratched." What's the story behind *black hole sun (scratched CD at UO / DEVIN KKENNY remix)* (2011/2020)?

DEVIN KENNY: It was made to sound like a CD that has been played so many times that it becomes scratched. It's *kind of* the original song. It's also *kind of* a remix, but it stutters and skips. It slows down and somewhat melts, along with other subtle alterations to the original.

When a CD gets scratched, usually songs will skip. One might hear certain weird, repetitive, or quiet noises. Or, the laser might have difficulty reading the film inside of the disk, and a CD will play up to a certain point; then stall, and start again . . .

FCN: When I first heard your version of "Black Hole Sun," I assumed that I was listening to a recording of a CD that had a scratch on the disc. As if you found a CD that was already scratched, like an *object trouvé,* and recorded it as a separate audio file that could be streamed. But that's incorrect . . . correct? How did you "scratch" it?

DK: I used Ableton Live—they did not sponsor me! But that's what I used. With that software, if you drop in pretty much any music in a common time signature that it can easily read, the software will determine how to visually organize the track. Ableton will say, OK, this is probably where the emphasis is, this is where the kick should be, this is where the snare is, etc. Then, it drops in little warp marker transients to represent how the rhythm is being organized. You can modify those transients to modify the rhythm or pulse of a song.

In certain ways, this warping technique has a relationship to another form of "scratching" that is analog, not digital. Take, for example, a turntable where you can beat juggle [manipulate two or more samples simultaneously] and change the rhythm of a record by scratching at particular rhythms or counterrhythms. The scratching done for *black hole sun (scratched CD at UO / DEVIN KKENNY remix)* is not quite that, but it's related. It's a "digital" approach.

FCN: Right. The term "digital" also makes me think of making something "with the fingers," which is also what you did. The warp-marker transients in Ableton that get yanked around are little inverted triangles displayed on a screen. One can pull on them with a cursor to snap them into different time-organizing grids or stretch them so they spill over the metrical grid. It's a physical process and not a purely conceptual activity.

DK: Yes. Exactly.

FCN: When listening to music through formats like a CD (or a WAV, MP3, cassette, record, etc.) and so on, we only hear "the song," not the CD. The media apparatus disappears into the background and becomes imperceptible to our ears. On the flip side, if the media is corrupted or not functioning correctly, like a CD with surface damage, we hear scratches and pops that stick out like a sore thumb. The materiality of the media disrupts the smooth transmission of "the song."

From my perspective, the scratched remix you made draws our attention to today's absence of physical media objects. Most of our everyday musical experiences are digitally generated and emerge from zeroes and ones, like everything else a computer reads. You can pop a CD-ROM out of a Walkman, drop it, scruff it up, subject it to sun damage, or what have you. But digital languages? They're opaque. It feels pretty wrong [laughter] . . . or at least counterintuitive to "scratch" a zero.

DK: I was interested in trying to use the thing wrong, especially in ways that counter the logic of tutorials that demonstrate how the software is supposed to work. I wanted to intervene. I didn't want to make something in a particular time signature either. You know how Ableton tutorials want you to take different kinds of indigenous music(s), and then force-feed them into common time [4/4], so you can DJ them at a party? [laughter]

So, what if I don't do that? Or, what if I just do it a little bit, but don't conform to the grid?

FCN: Absolutely! I also share a similar attitude towards "the grid."

You research alternative organizing principles and the people who employ them quite often in your work. Messing with grids that organize musical time also played a big role in another, more recent project of yours, the full-length album *los giros de la siguiente* (2019). Can you tell me some more about this project?

DK: When I moved from Brooklyn to Houston, I was working on a project that involved looking at the archive of DJ Screw's remixes, which was collected by the University of Houston.

DJ Screw (b. Houston, 1971–2000) was a DJ and producer who basically created a subgenre of hip-hop called Chopped and Screwed. Typically using soul, rap, and other kinds of music samples, he would play back selections at a slower tempo. For example, you take a 90 BPM [beats per minute] song and play it around 60 BPM, or even slower, which syncopates everything. In this process there will be some "scratching," which DJs call "chopping," and the original sample will also be pitched down and warped. It's pretty much a low and slow version of particular songs.

There's also a subgenre of *cumbia* that emerged simultaneously in Monterrey, Mexico, called *cumbia rebajado*. This style basically involves playing a cumbia record at a slower tempo. Apparently, this technique came about by way of accident. People were playing records, being DJs, throwing parties, and going long into the night. The batteries on the tape deck would get depleted and then the records would start warping and playing low and slow. People loved it.

FCN: After hearing your album, I listened to some cumbia rebajado and that "low and slow" effect sounds positively psychedelic and out of this world, almost gothic. It was different from any cumbia that I've heard before. As an overall genre of dance music, cumbia is remarkably flexible. It's got that distinctly relaxed rhythm built from a mix of indigenous Colombian folk rhythms, African slave rhythms, and Spanish colonialist influence. It's infectious party music too, and one of the most widespread musical genres of Latin American origin. It spread up to Mexico in the 1980s, where it's still immensely popular, especially with the working class. Taking a cumbia record and playing it at less revolutions per minute [RPM] warps and intensifies the poly-centered pulse of the music's "lower-class" rhythms. It projects all the angst that late twentieth-century Western art music hoped to achieve yet makes that angst joyously irrelevant. [laughter]

Really, both styles do, and the connections you make between them in the album emphasize this. Chopped and Screwed and cumbia rebajado propose an alternative approach to how we typically conceive of instruments. The actual "playing" of the songs is the instrument. It's more of a music theory of the interface perhaps . . .

Dora Budor

X ZHU-NOWELL: Middling with interfaces and expanding the idea of what an instrument is reminds me of your work, Dora. What is an instrument to you?

DORA BUDOR: When Devin was talking about his scratched remixes, it reminded me of William

Basinski's "Disintegration Loops" (2002–3), which I really love. What captivates me about those is how there are different ways of embedding temporalities into sound—weathering of the magnetic tape when it's played over and over until it physically dissolves, and then its reanimation through the digital encoding and reproduction, making it stable and present again.

The ability of sound to work simultaneously and across different timelines interests me, as it allows for it to function both as a carrier and a trigger. Sound is always already relational, as it leaks in between things, spaces, and persons, but in my work those relations often route to the hidden, invisible, or a something that operates at a different scale. For example, when I did a show at Kunsthalle Basel, "I am Gong" (2019), I used all the infrastructural hollows that accommodate electricity, air-conditioning, water, heat, media channels, and other utilities in the building. These cavities behind the walls, ceilings, and floors became resonant surfaces and hosted a 14-channel sound piece, which modulated in real time by reacting to the construction process nearby.

One can think about the entire building becoming an instrument. In this case, our sensing becomes detoured, allowing for the unity of the infrastructural system behind the white cube to take over, which in return undoes the architectural space. Also, that museum is such a beautiful, majestic space with six-meter ceilings, and as the unrealized architect that I am . . .

XZ-N, DK, FCN, CB: [laughter]

DB: I had to break it [the building] down, in a way. Sound has the capability to completely rephrase and restructure the architecture of the space. All that is to say that an instrument, in the case of my work, can be a building. Or, it can be something that reacts to visitors' movements, like in *Adaptation of an Instrument* (2016), which reacts to visitors' movements. In this environment, the sensitivity builds up according to visitors' activity, which triggers a light system in its walls that is modeled after neurological pathways in the human body. So, you're looking at the impulses, the excitation and inhibition—"music" of the vision being generated while looking at the environment . . .

FCN: Have you ever come across the term "organology" by any chance?

DB: Organ as in instrument?

FCN: Yes! I mean, it's a little niche. [laughter] The term is pretty much just *organ* [i.e. instrument] + *ology*. It puts forth the idea that instruments are not just mere aiding devices, but also express the ways in which we conceive of the objects of our study, which is particularly useful for music and other kinds of temporal objects. You know, notoriously ephemeral things . . .

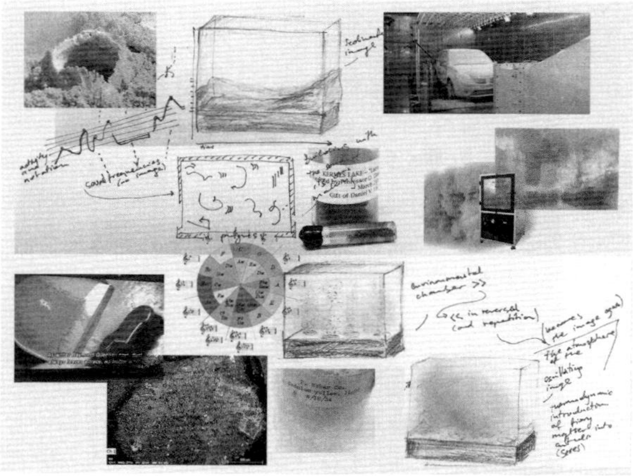

Dora Budor, preparatory sketch for *Origin (. . .)* series

DB: I'm using the idea of instruments to apply to non-instruments. To see how they can function as instruments. How can one *play* a space? How can you, by using sound or similar affective strategies, narrate a different experience than the prescripted, normative one. It opens up some alternative, minor ways of experiencing the built environment. It goes beyond the major language of architecture that is a product of a culture increasingly dominated by symbolic capital.

There is also so much about how one's experience and perception of a space shifts depending on . . . I mean, if I play Alanis Morissette's "Hand in My Pocket" and walk down the street, my reading of that street and its social dynamics might be completely different! The sensuality and affective potential of sound is very powerful.

DK: I love Alanis.

FCN: I love that. You're testing the limits. This is a much different approach than, say, associating instruments with certain aesthetic commitments. Such as tasking a harp to play angelic sounds.

DB: I also think of instruments from a scientific perspective. It's something you can play and produce, but it's also something that measures and makes things visible, like anemometers, seismographs, all types of different sensors. I think of an instrument as something in between things that you can use for different ends and means. I've tried to use it from both sides, as a measuring mechanism and a producing apparatus.

FCN: I'm very drawn to the way that you consider sensitivity of your audience and explore the relationships between measuring apparatuses and our intimate perceptions of space. The question "How can one *play* a space?" is quite provocative.

I know that you often create environments in your practice. Sometimes they are contained or highly controlled, and other times an environment will be intertwined with a larger, more nebulous surrounding environment outside of a work's immediate exhibition context. Can you talk a bit more about environments factor into your practice?

DB: It varies from project to project. For the exhibition at Kunsthalle Basel, I made a series of pieces titled *Origin* that employ the technology of the environmental test chambers used in factories to test and accelerate the effects of time and weather on manufactured objects. The "weathering" comes from snowstorms, seismic movements, radiation, thermal shock, hail, rain; these are simulated in the chambers to test the limits of an object. When does this thing start to rust, fall apart, and so on?

Inside the chambers are dusts and pigments that are injected into the environment, and it gets filled with clouds and pollution, but the score for the piece comes from outside of the museum. So, it depends on the surrounding environment. If there is a construction site or a development nearby, a sensor from there will transmit the sound frequency in real time back to it. Therefore, the piece changes throughout the exhibition—it is constantly forming into something, and then unforming and dissipating. Somehow, every time that I have shown it, the museum has been in the vicinity of a construction site. Obviously, this says a lot about the neoliberal strategies of urban development, and prime real estate in which cultural institutions are positioned.

There are three different pieces in this series, and in each the hues are based on one of J.M.W. Turner's paintings (1775–1851). In the context of Western art history, these paintings are early examples of an artist focusing on the atmosphere as something changing, a thermodynamic introduction of fiery matter into culture. One of the

Left: Dora Budor, Installation view, "I am Gong," Kunsthalle Basel, 2019. *Origin I (A Stag Drinking)*, 2019; *Origin II (Burning of the Houses)*, 2019; *Origin III (Snow Storm)*, 2019. Right: Dora Budor, *Origin III (Snow Storm)*, 2019. Installation view, "I am Gong," Kunsthalle Basel. Custom environmental chamber (reactive electronic system, compressor, valves, 3D printed elements, aluminum, acrylic, LED light, glass, wood, paint), organic and synthetic pigments, diatomaceous earth, FX dust, felt

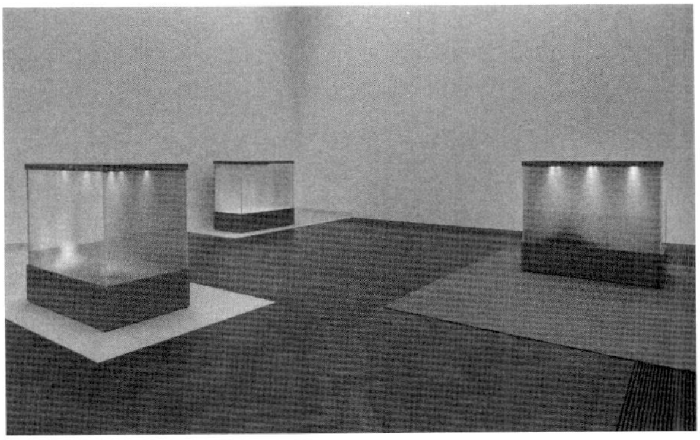

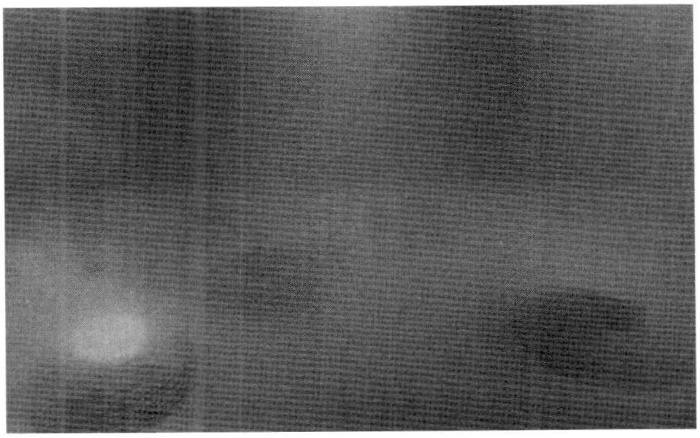

chambers is in reference to a Turner Petworth sunset painting, which was intensely yellow due to the presence of all the volcanic ash and gas in the sky. It kind of looked like LA after the fires, because of all the particles refracting the light, making a yellow-orange atmosphere. That piece connects distant temporalities, but also reflects on a historical moment captured in a painting, as something that is then preserved in a museum and can never be changed. It creates an evolving image-form that is continuously building up, and then falling apart.

XZ-N: I'm very interested in what you said about hue, which I also associate with RGB color lighting. It has a digital correlation and a physical manifestation. What do you think about the line between the artificial and natural when it comes to environments?

DB: Nothing has not been unaffected in one way or the other by things that humans have created. That makes nothing unnatural or artificial, as everything exists in an entangled way.

XZ-N: Right. The definition of natural is very different now, it's not such a Romantic idea.

Cecilia Bengolea

XZ-N: Cecilia, your practice utilizes movement and dance in ways that intimately intertwine bodily and environmental energies. Climate and weather, particularly rain and bodies of water, are just as dynamic as the choreography. The environment connects beings together in these liquid states, or as you call it, "a state of greater liquidity."

I'm specifically thinking about your videos *Lightning Dance* (2017) and *Synchronized Serpent* (2018). The former captured a series of collaborative Dancehall improvisations that took place on the streets of Spanish Town, Jamaica, during the flood season. The latter, *Synchronized Serpent*, was created in collaboration with the Jamaican national synchronized swimming team. An aerial camera captured the dancers improvising in the ocean from above and their bodies become entangled in the

Cecilia Bengolea, *Lightning Dance*, 2017

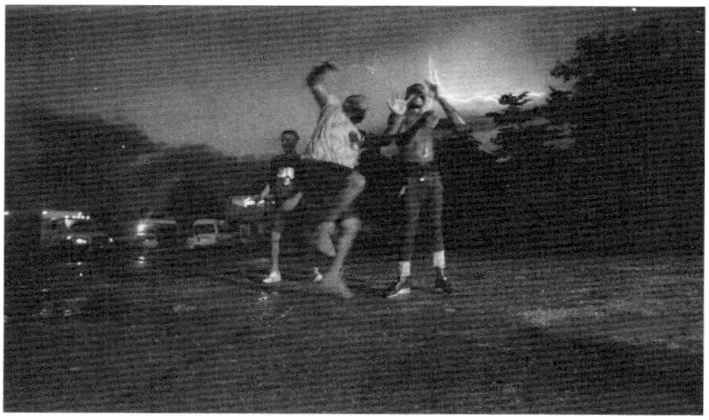

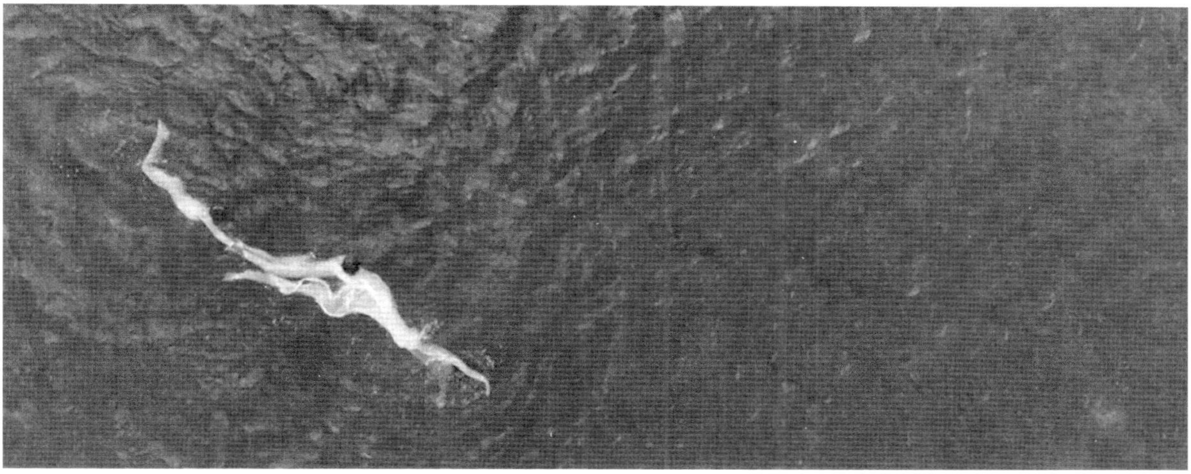

Cecilia Bengolea, *Synchronized Serpent*, 2018

sea and gradually aggregate into a larger, collective body. Can you talk more about how this work came into being?

CECILIA BENGOLEA: I was doing a residency in Jamaica at the TBA21 foundation and nearby there was a pool where the national synchronized swimming team trains. I started talking about dance with the coach, Olga, who happened to be an Olympic gold medalist from Russia who married a Jamaican guy and stayed there. She started the team with at least fourteen swimmers who were quite young, between six and fourteen years old.

They had never done choreography outside of the swimming pool, but they were all looking forward to performing in the ocean. I wanted them to embrace the fluidity of being in the water, their movements, and improvising together. The team hadn't done much improvising before together, and I wanted the girls to break from the very precise ways that they had been trained to move. It was important to get them to embrace the ways that their bodies are driven by a physical intelligence of their own, so we spent time on dry land thinking up imaginary, mythical creatures together before we improvised and began making choreography for the ocean.

I brought the dancers some different books that detailed mythological creatures, such as [Jorge] Luis Borges's *The Book of Imaginary Beings,* the Qur'an, among others. Some stories resonated more than others.

They also shared some of their own myths with me to get the creative process started. One creature particularly resonated with the group, which the title of the video reflects. *Synchronized Serpent* is a nod to a Jamaican myth about a half-serpent, half-bird creature that has four wings and sings very loud songs. As some of the girls were describing this being to me, they also did little dances to animate and personify how it moved in the air and on the earth. From this, we began to improvise and create other imaginary beings that were part earth, part air, and part water. Many were complex and fantastic, and I picked a couple for them to focus on, which they transferred to the pool. They were also singing in the swimming pool, which was really nice.

XZ-N: Can you tell me more about the different mythological characters and hybrid beings that the dancers came up with?

CB: One of the girls that was featured as a soloist in the video was working on a *poisson volant* [flying fish] character that lived in the trees and was scared of the water. It was half fish and half . . . bird! I guess, because it's a flying fish. [laughter] She personified the being in a complex way with her body that didn't separate action from thought. Since the character had a fear of water, she performed some very rigid, straight movements because that's how she imagined a flying fish would react to touching water. Out of fear, this being would freeze up and become stiff. In the video you can see a solo dancer performing rigid, half-quarter turns at an angle. This is the girl as the flying fish, whose body was rejecting being in this environment, the water.

Another dancer was also making very symmetrical shapes like triangles with her legs and moving in circles. Her mythological being was an *aspidochelone,* a huge, fabulous mother turtle from the Medieval Bestiary [*Physiologus* (2nd c. CE)]. Her back was huge, like an island, and covered with crustaceans. She hosts many others and collaborates with other species, so she has to move very slow, because . . . well, she's a turtle . . .

XZ-N, FCN, DK: [laughter]

XZ-N: It's wonderful to hear about the specificity behind all the movements. From an aerial perspective, the movements can look somewhat abstract.

CB: Yeah, it maybe looks abstract, but it comes from some thick description and careful imagining. The children were also adept at making decisions on the spot, which was nice and quite different from how they are trained to make figures for their competitions. This time, it was not so imposed. I didn't want them to do so much of the typical synchronized choreography.

FCN: How did the coach teach them choreography? Did she demonstrate figures or make diagrams? Are they working from some kind of shared movement vocabulary?

CB: Yes, they have a vocabulary, but I didn't see them write down anything. It was the first time they were doing synchronized swimming in the open ocean, so we improvised quite a lot. Specifically, we improvised with the sensation of being connected. They were also hearing different kinds of things that they responded to in the pool, because when you are floating in this environment you can hear different sounds coming from below, depending on the depth you are at.

We would practice in the swimming pool in preparation for shooting specific sequences. In the ocean, they were really playing with bodily imagination and perspective, and it wasn't that difficult for them.

XZ-N: Can you say more about how you and your collaborators worked in this environment, in the ocean? In a synchronized swimming competition, I would assume that the swimmers are able to listen and count to the music, so they can flip and turn together and things of that sort. How did the dancers synchronize together in the open ocean, where they don't have any kind of musical structure to rely on?

CB: Yeah, that's true! That was quite mysterious to me, because in the swimming pool they had a speaker under the water. When Olga, the coach, initially showed me their choreography she would shout out "1-2-3-4" and "tra-la-la" loud enough so the girls could hear it underwater and react. However, they didn't have anyone counting or a speaker in the ocean . . . So, I guess they were counting in their head, like all dancers do.

Through repetition they were also relieved of having to rely on call-and-response mechanisms that direct their minds to tell their bodies what to do. My work contemplates liquidity as a form of constant evolution and transformation, so I wanted to keep this fluidity present in the video. By shooting the dancers from above I could synchronize and recompose their individual bodies within a state of greater liquidity. The dancers, the water, and all the other elements and organisms in the environment conformed to a collective body that was very much alive, together. I like to think of my works as being alive.

CONTRIBUTORS

Cecilia Bengolea is a multidisciplinary Argentinian artist who works in video, sculpture, installation and performance. She uses dance as a tool and a medium for radical empathy and emotional exchange. Through collaborations with artists, performers, DJs and dancers, Bengolea develops a broad artistry where she sees movement, dance and performance as animated sculpture.

Dora Budor is a Croatian-born, New York–based artist. Using installation, environment, sound, and sculpture, her works are constructed as complex systems in which each element exists in interdependency, and responds to the context and site of the exhibition, reacting and negotiating their initial conditions.

Lou Cantor is a Berlin-based artist collective founded in 2011 whose work is grounded in intersubjectivity. Cantor's practice explores the polysemic minefield of contemporary communication, where medium, message, and meaning constantly fold back into each other.

Constant Dullaart's practice reflects on the broad cultural and social effects of communication and image processing technologies, and ranges from performatively distributing artificial social capital on social media to completing a staff-pick Kickstarter campaign for a hardware start-up called DulltechTM. His work has taken the form of websites, performances, routers, installations, startups, armies, and manipulated found images, and frequently juxtapozes or consolidates technically dichotomized presentation realms.

Hal Foster is Townsend Martin Professor of Art and Archaeology at Princeton University. A co-editor of *October* magazine and books, he is the editor of *The Anti-Aesthetic* (1983), and the author of *Design and Crime* (2002), *Recodings: Art, Spectacle, Cultural Politics* (1985), *The Return of the Real* (1996), *Compulsive Beauty* (1993), and *The Art-Architecture Complex* (2011).

Kevin Gotkin is a Visiting Assistant Professor of Media, Culture, & Communication at NYU. From 2016 to 2019, he was Co-founder of Disability/Arts/NYC with Simi Linton.

Camille Henrot is a French artist whose practice moves between film, painting, drawing, sculpture, and installation. She references self-help, online second-hand marketplaces, cultural anthropology, literature, psychoanalysis, and social media to question what it means to be at once a private individual and a global subject.

Sun-ha Hong's research interrogates the social life of technological fantasies, with an emphasis on AI and data-driven surveillance. He is the author of *Technologies of Speculation: The Limits of Knowledge in a Data-Driven Society* (NYU Press, 2020), and is currently Assistant Professor of Communication at Simon Fraser University.

Tobias Kaspar is an artist based in Riga and Zurich. He is interested in cultural transfer processes of styles and fashions and tracing global trajectories. The mechanisms of fashion—its dazzling power of seduction, its ability to create the desire to be someone else, and its capacity to embody or simulate various identities—are a key frame of reference for his practice.

Devin Kenny is a New York–based artist who works across various mediums, including sculpture, video, photography, painting, design, and music, and also writes and organizes exhibitions. See his various output at www.linktr.ee/devinkkenny.

Agnieszka Kurant is an artist whose work investigates collective intelligence, nonhuman intelligences (from microbial to artificial intelligence), and the exploitations of social capital under surveillance capitalism. She explores the transformations of the human and the future of labor and creativity in the twenty-first century.

Lynn Hershman Leeson is an artist and filmmaker whose work investigates the relationship between humans, technology, identity, and surveillance, as well the use of media as a tool of empowerment against censorship and political repression. Over the last fifty years, she has made pioneering contributions to the fields of photography, video, film, performance, artificial intelligence, bio-art, installation, interactive, and net-based media art.

John Miller is an artist, writer, and musician based in New York and Berlin. In 2020, the Schinkel Pavillon in Berlin featured an overview of his work titled "An Elixir of Immortality." In June 2021, the Kunsthalle Bielefeld exhibited his *Middle of the Day* photographs with his video and PowerPoint works. Miller is a Professor of Professional Practice in Barnard College's Art History Department.

Frederick Cruz Nowell is a musicologist and PhD candidate at Cornell University. He focuses on the intellectual history of aesthetics from a cross-disciplinary perspective that draws upon discourses in historical musicology, modern and contemporary art history, the history of music theory, and transgender history/the history of gender variance. His current research investigates the development of abstraction in the context(s) of twentieth-century modernism.

Samantha Ozer is a curator and writer based between London and New York. She is currently a content researcher and writer for the Poetic Justice group at the MIT Media Lab.

Aleksandra Przegalińska received her doctorate in the field of philosophy of artificial intelligence at the Institute of Philosophy of the University of Warsaw. She is currently Associate Professor at the Department of Management in Digital and Networked Societies at the Leon Kozminski Academy and the Leader of AI in Management Program, and in fall of 2021 will be a Senior Research Fellow at the Labor and Worklife Program at Harvard University. She is interested in the development of new technologies, natural language processing, humanoid artificial intelligence, social robots and wearable technologies. With Dariusz Jemielniak, she is the author of *Collaborative Society* (MIT Press, 2020).

Farid Rakun trained as an architect (B.Arch from Universitas Indonesia and M.Arch from Cranbrook Academy of Art) but wears different hats, depending on who is asking. A visiting lecturer in the Architecture Department of Universitas Indonesia, he is also a part of the artists' collective ruangrupa, with whom he is currently serving as a collective Artistic Director for documenta fifteen in 2022.

Tiana Reid is a scholar and writer from Toronto. She has written for many publications, including *Art in America*, *Bookforum*, *Frieze*, the *New York Review of Books*, the *Paris Review*, and *Teen Vogue*.

Patrick Urs Riechert is a researcher exploring the intersections of politics, cultures, and the digital. He is particularly interested in the history and mechanics of governing, as well as processes of subjectivation.

Isabel de Sena is a Berlin-based independent curator and currently a PhD candidate at the University of Reading School of Art. Her current research explores the use of science/speculative fiction within curatorial practice, focusing on non-Western and nonhuman perspectives on earthly survival that undermine the mythical distinction between self and non-self in relation to geobiological/social environments.

Jenna Sutela works with words, sounds, and other living media, such as *Bacillus subtilis nattō* bacteria and the "many-headed" slime mold *Physarum polycephalum*. Her audiovisual pieces, sculptures, and performances seek to identify and react to precarious social and material moments, often in relation to technology. Sutela's work has been presented at museums and art contexts internationally, including Guggenheim Bilbao, Moderna Museet, and Serpentine Galleries. She is a Visiting Artist at the MIT Center for Art, Science & Technology (CAST) in 2019–21.

Elena Vogman is a scholar of comparative literature and media and independent curator. She is currently Visiting Assistant Professor at New York University in Shanghai. From June 2021 she will be a Freigeist Fellow of the Volkswagen Stiftung at the Bauhaus University Weimar with a research project entitled "Madness, Media, Milieus. Reconfiguring the Humanities in Postwar Europe."

Emily Watlington is a critic, curator, and assistant editor at *Art in America*. A Fulbright scholar with a master's degree from MIT, she often writes about topics including art, design, disability, and feminism. She is based between Cambridge, MA, and New York.

X Zhu-Nowell is an Assistant Curator at the Guggenheim Museum in New York. Prior to joining the Guggenheim Museum in 2014, they were part of the curatorial team for the 14th Istanbul Biennial. They hold a master's degree from MIT. Since 2018, X Zhu-Nowell has collaborated with the musicologist Frederick Cruz Nowell as Passing Fancy, a curatorial and artistic research duo.

Relative Intimacies
Intersubjectivity Vol. III

Published by Sternberg Press

Editors:
Lou Cantor, Emily Watlington

Proofreading:
Grace Sparapani

Design:
Katy Nelson for Joseph Logan Design

Printing and binding:
Zakład Poligraficzny SINDRUK
Firmowa 12, 45-594 Opole, Poland

Paper:
135gsm GardaPat Klassica

Note on the type:
This book is typeset in Eliza and Theinhardt. Designed by Pawel Wolowitsch for Camelot Typefaces, Eliza explores the relationship between the human and the machine by merging calligraphic strokes with the square form of the pixel. Eliza takes its name from the pioneering natural language processing program that Joseph Weizenbaum developed at MIT in the 1960s. Designed by François Rappo for Optimo Type Foundry, Theinhardt is a contemporary take on early grotesque designs.

ISBN 978-3-95679-625-8

© 2021 the authors, the editors, Sternberg Press
All rights reserved, including the right of reproduction in whole or in part in any form.

Distributed by The MIT Press, Art Data, and Les presses du réel

Sternberg Press
71–75 Shelton Street
London WC2H 9JQ
www.sternberg-press.com

Cover image: Lou Cantor, courtesy of the artist. 6, 11: Lou Cantor, courtesy of the artist. 22, 25–26, 28, 31: Agnieszka Kurant, courtesy of the artist and Tanya Bonakdar Gallery (New York/Los Angeles). 52–57: Jenna Sutela, courtesy of the artist. 67: *Zoolander*, Ben Stiller, 2001. 67: *Wall Street*, Oliver Stone, 1987. 78, 82, 85–86: Lou Cantor, courtesy of the artist. 92: Camille Henrot, courtesy of the artist, kamel mennour (Paris/London), König Galerie (Berlin/London/Seoul), and Metro Pictures, New York. 103–104, 107, 110, 110: Lynn Hershman, courtesy of the artist. 112–118: Constant Dullaart, courtesy of the artist. 120: Wikimedia Commons. 123: Archives Marie-Dominique Guibal, courtesy L'Arachnéen, reproduction Anaïs Masson. 123: René Laloux. 126: Bridget Riley. 126: Leigh Hendricks. 128: Masao Komura and Kunio Yamanaka. 131: Frederick Rowland Emett. 153: Camille Henrot, courtesy of the artist, kamel mennour (Paris/London), König Galerie (Berlin/London/Seoul), and Metro Pictures, New York. 162–169: Tobias Kaspar, Marie Matusz, Ivan Mitrović, and Cassidy Toner, courtesy of the artists. 174: Dora Budor, courtesy of the artist. 175: Dora Budor, courtesy of the artist and Philipp Hänger / Kunsthalle Basel. 175: Dora Budor, courtesy of the artist and Gina Folly / Kunsthalle Basel. 176–177: Cecilia Bengolea, courtesy of the artist. 184: Lou Cantor, courtesy of the artist

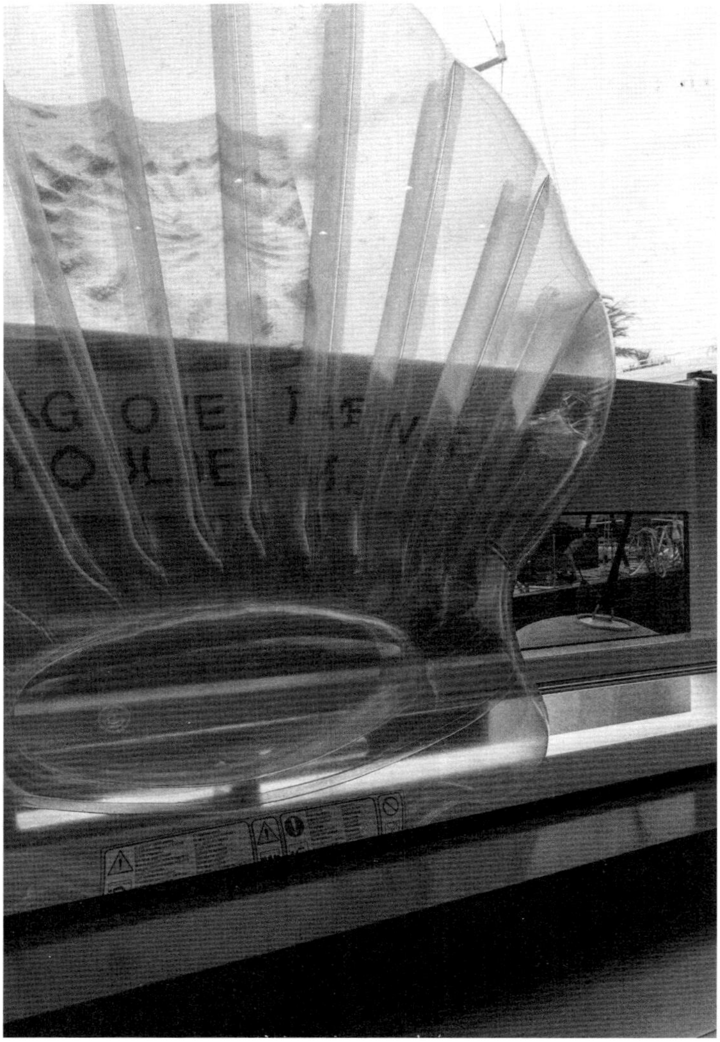